CLIMATE CHANGE AND FUTURE JUSTICE

Climate change creates unprecedented problems of intergenerational justice. What do members of the current generation owe to future generations in virtue of the contribution they are making to climate change?

Providing important new insights within the theoretical framework of political liberalism, *Climate Change and Future Justice* presents arguments in three key areas:

- **Mitigation**: the current generation ought to adopt a strong precautionary principle in formulating climate change policy in order to minimise the risks of serious harm from climate change imposed on future generations.
- **Adaptation**: the current generation ought to create a fund to which members of future generations may apply for compensation if the risks of climate change harm imposed on them by the current generation ripen into harms.
- **Triage**: future generations ought to keep alive hope for a return to the framework of justice for social cooperation between future people less burdened by climate change.

This work presents agenda-setting applications of important principles of democratic equality to the most serious set of political challenges ever faced by human society. It should be required reading for political theorists and environmental philosophers.

Catriona McKinnon is Professor of Political Theory at the University of Reading. She has published widely on many topics in contemporary political theory, including toleration, political liberalism, equality, and cosmopolitanism.

Routledge issues in contemporary political theory

This series engages with the most significant issues in contemporary political theory. Each text is written by a leading scholar and provides a short, accessible and stimulating overview of the issue for advanced undergraduates and graduate students of political theory. As well as providing a survey of the field, the books also contain original and groundbreaking thinking which will drive forward debates on these key issues.

1. **Sovereignty and Political Theory**
   Indigeneity and the limits of the political
   Karena Shaw

2. **Fascism and Political Theory**
   Critical perspectives on Fascist ideology
   Daniel Woodley

3. **Climate Change and Future Justice**
   Precaution, compensation, and triage
   Catriona McKinnon

CLIMATE CHANGE AND FUTURE JUSTICE

Precaution, compensation, and triage

Catriona McKinnon

Routledge
Taylor & Francis Group
LONDON AND NEW YORK

First published 2012
by Routledge
4 Park Square, Milton Park, Abingdon, Oxon OX14 4RN

Simultaneously published in the USA and Canada
by Routledge

605 Third Avenue, New York, NY 10017

Routledge is an imprint of the Taylor & Francis Group, an informa business

British Library Cataloguing in Publication Data
A catalogue record for this book is available from the British Library

Library of Congress Cataloging in Publication Data

ISBN 13: 978-0-415-46124-5 (hbk)
ISBN 13: 978-0-415-46125-2 (pbk)
ISBN 13: 978-0-203-80220-5 (ebk)

Typeset in Bembo
by Taylor & Francis books

For my children, Caelan and Bria

CONTENTS

ACKNOWLEDGEMENTS

In a number of ways, this has been a difficult book to write. The subject matter is relentlessly depressing. At present, there is very little ground for hope that we, the current generation, will rise to the political challenges created by climate change, and the time available to us to mitigate in any significant way is fast running out. Reflecting and writing on Rawls' just savings principle, or the differences between strict and fault liability, or the nature of principles of triage, etc., has much of the time felt like fiddling while Rome burns. However, to have been reflecting and writing on anything else would have felt even less appropriate. The book was also difficult to write for (far nicer) personal reasons. The process was punctuated by the conception and arrival of my two beautiful children, Caelan and Bria, and by extended periods of maternity leave devoted exclusively to them. As anyone with young children knows, as life-enhancing and joyful as it is to have them around, they do not conduce to the concentration needed to write a book in political theory. It is only because of the heroic efforts of my husband, Matt Pittori, by sacrificing his rest, to ensure that I got sufficient sleep to work that this book exists at all. He often 'did the nights' with Caelan and Bria when they were restless babies, and then 'did the days' too, finding the energy to run, sing and build sandcastles with them while I sat with books and a keyboard. Our children could not have a better father. I am hugely proud of him, and love him more than ever.

Bits of this book have done the rounds over the last few years, and I have had much helpful and encouraging feedback from various audiences. Comments from the following people in particular prompted some major developments and revisions to early drafts: Harry Brighouse, Gideon Calder, John Horton, Cécile Laborde, Ed Page, Nigel Pleasants, Andrew Williams, Victor Tadros, and Jo Wolff. I would also like to thank Routledge's anonymous reader of the penultimate draft, whose comments forced me to not to stray too far from the theme.

For the duration of the book's gestation I have been very lucky to be employed in the Department of Politics and International Relations at the University of Reading. As my colleague Alan Cromartie puts it, there's no politics in the Politics Department, making it a civilised and healthy place to work. Thanks are due to Alan, and my other colleagues, for their supportiveness and sanity. I would also like to thank my Head of School, Philip Giddings, for his calm and thoughtful handling of my periods of maternity leave, ensuring that these times were spent as they should be.

I have enjoyed two periods of extended funded research leave to work on the topics explored in this book. In 2006 I took up a Research Fellowship funded by the Leverhulme Trust (grant no.: RF/2006/0303), and in 2007 I took up a Senior Research Fellowship funded by the British Academy and the Leverhulme Trust (grant no.: SRF/2006/12). I am grateful to both bodies for enabling these sustained periods of thinking and writing. In connection with the Senior Research Fellowship, I would particularly like to thank Ken Emond at the British Academy for cheerfully managing the bureaucracy of two requests to interrupt the Fellowship for maternity leave.

Thanks are also due to my editor, Craig Fowlie, at Routledge for his advice and guidance, and for his sympathetic responses to my (repeated) requests for extensions to the deadlines. Nicola Parkin at Routledge has also been very helpful.

Over the last few years I have found that my friendships have diminished in number but increased in depth. I know that however long it is since I have seen them, the following people are always there, and that we remain as we were the last time we met: Havi Carel, Karen Chung, Cécile Laborde, and Nigel Pleasants. There are far fewer late nights and there is much less booze now than there used to be, and far more children, but just as much love.

Finally, this book is dedicated to my son, Caelan, and my daughter, Bria. When I started thinking about climate change and future generations I was not a parent, and my concerns were abstract. Having children has made doing the research into the likely severity and future impacts of climate change on the planet sometimes acutely painful, but returning to their funny faces and tiny worlds at the end of a working day has been a delight. It is because I cannot stand the thought of what they might have to watch their own children face as a result of my generation's greed and cowardice that this book was completed.

Bournemouth
April 2011

1

INTRODUCTION

Climate change is not a new phenomenon. The last three million years of the Earth's history have been punctuated by Ice Ages, and during most of the past five hundred million years the Earth was much warmer than it is today, and probably completely free of ice sheets. Major shifts in the climate, and large fluctuations in global mean temperature with all this entails for weather patterns, extinction events, and the direction of evolution of species, are a normal, and sometimes predictable, feature of the planet's history. Past climate change has had natural causes: changes in incoming solar radiation (for example, as a result of solar flares), changes in the fraction of solar radiation reflected back into space (for example, as a result of changes to the planet's albedo), and alterations in the longwave energy reflected back into space (for example, as a result of clouds of volcanic ash which have a cooling effect).[1] The climate change we are currently experiencing differs from past shifts in two crucial respects: it is man-made, and it is happening more speedily than at any time in the last fifty million years.[2] If current climate change were not anthropogenic, its time scale and potential effects on life on Earth would still make it the biggest, most difficult global political challenge ever faced by humanity. That the causes of current climate change are anthropogenic mean that there is an increased – although annually diminishing – chance that we can mount adaptation strategies to deal with the effects of climate change already in the pipeline, and take action to prevent and mitigate climate change further down the line.[3]

The following features of climate change and its effects combine to make doing something about it hugely – and perhaps uniquely – challenging.

- The seriousness of the problem. In worst-case scenarios of runaway climate change, *Homo sapiens* could go extinct, and even in best-case scenarios it is likely that many other species will become extinct in virtue of being less able than human beings to adapt to quickly changing environmental conditions.

- The immediacy of the problem. Optimistic authorities estimate that we have a period of ten years in which to make drastic cuts in greenhouse gas emissions, in particular, cuts in carbon dioxide (CO_2) to prevent concentrations of it exceeding 400 parts per million (ppm). Beyond this concentration the probability of preventing a global surface temperature rise of 2°C above pre-industrial levels is severely restricted, and climate change may become irreversible. In March 2004 the concentration of CO_2 stood at 379ppm. Less optimistic authorities claim that the point of no return for climate change is almost upon us.
- The duration of the problem. The most common greenhouse gas, CO_2, remains in the atmosphere for around a century: *per impossibile*, achieving zero emissions of CO_2 tomorrow would likely not affect the climate until 2105.
- The complexity of the problem. Consensus in the scientific community on the existence of climate change, and its roots in anthropogenic greenhouse gas emissions, masks much disagreement about the mechanisms and processes of climate change, and its likely impacts (on, for example, ecosystems, weather systems, ice sheets, the thermohaline ocean circulation).[4] Our scientific understanding of climate change lags behind the phenomenon.
- The global scope of the problem. Emissions do not remain at their source but rather, spread evenly throughout the atmosphere across the globe. Climate change creates a collective action problem for the global community on a scale never before encountered.

The climate change we are experiencing requires swift, far-ranging, globally coordinated and probably very unpopular political action. Depending on the action taken in the next ten to twenty years, the current generation will either stand as the cohort who pulled humanity back from the brink of environmental disaster – possibly even from extinction – or as the cohort who missed the last big opportunity to do so. The current generation did not create the climate change problem (although many of us are doing much to exacerbate it), and are horribly unlucky in having been born at a time in human history when the science that reveals the scale and potential devastation of anthropogenically caused climate change is maturing just in time to show how little time we have left to do something about it. The intractability of the climate change problem, and its epic – indeed, melodramatic – dimensions can paralyse thinking about what must be done, and how to do it. Thinking through the political problems of climate change and – worse – the consequences of doing nothing can have an unreal and nightmarish quality.

This book presents some arguments in political theory for principles to guide policy making with respect to climate change. The focus is on securing justice for future generations, who will be the worst affected by climate change, and the inspiration for the approach taken is the work of John Rawls. The urgency of the political challenges of climate change, and the complexities of framing and implementing policies fit to meet them, renders the ideals of pure political theory puny tools. Nevertheless, this book is written with the conviction that even if – as is likely – these ideals have no influence on policy, it still matters enormously that we know what they are, and why they are justified. Justice is a matter of trying to realise ideals, and its scope extends to the problems of climate change, however enormous and frightening they are. It is part of our humanity to think in terms of justice, and we should not abandon that even if the grounds for hope that we can achieve it disappear.

Chapter 2 elaborates aspects of the theoretical framework to be adopted in the book, and the final section of this chapter gives an outline of the arguments made throughout and how they proceed. For now, let me turn to the science and potential impacts of climate change, and to existing policy frameworks to address it.

Science and impacts

Current climate change is a result of the 'greenhouse effect'. This is a process whereby infrared radiation emitted from the surface of the Earth – which has been warmed by the Sun – is absorbed and re-emitted back towards the lower atmosphere and the Earth by greenhouse gas molecules and water vapour in clouds. The primary greenhouse gases are water vapour (H_2O), carbon dioxide (CO_2), methane (CH_4), and nitrous oxide (N_2O). Without the warming of the natural greenhouse effect, the average temperature at the surface of the Earth would be below zero degrees, and life on Earth as we know it would not be possible.[5] However, this natural greenhouse effect has been enhanced and intensified by human activity creating a higher atmospheric concentration of greenhouse gases, and of CO_2 in particular, which formed 77 per cent of total anthropogenic greenhouse gas emissions in 2004.[6] Overall, greenhouse gas emissions increased by 70 per cent between 1970 and 2004, and CO_2 by 80 per cent.[7] The study of ice cores – in which bubbles of ancient air are trapped and can be analysed – shows that the current concentration of CO_2 and CH_4 in the atmosphere far exceeds the natural range of concentrations for the last 650,000 years.[8] For example, in (pre-industrial) 1750 the concentration of CO_2 in the atmosphere was 280ppm, rising to 380ppm in 2006.[9] Corresponding to the increased concentration of greenhouse gases such as CO_2 are other indicators of climate change: for example, global mean surface temperature has increased $0.74° \pm 0.18$ over the period 1906–2005; global mean sea level rose at an average annual rate of 1–2mm in the twentieth century, and 3mm in the decade 1993–2003 (sea water expands as it heats up); total global snow cover has decreased by 10 per cent since the 1960s; and the growing season has lengthened by one to four days per decade in the last fifty years in the northern hemisphere.[10]

According to the Intergovernmental Panel on Climate Change (IPCC),[11] most of the observed increases in global mean surface temperature since the mid-twentieth century are very likely (i.e. have a 90 per cent probability) to have been caused by anthropogenic greenhouse gases.[12] When temperature rises 1900–2000 are plotted as predicted by models using only natural causes, and those using natural and anthropogenic causes, the latter correspond far more closely to actual observed temperature rises than do the former.[13] Human beings are the cause of current climate change, and industrialisation (with associated increased demands for energy from coal-based sources, and increased use of oil-based transport systems), agriculture (e.g. methane emissions from livestock) and changes in land use (e.g. large-scale deforestation, which deprives the planet of carbon sinks) are the key activities.[14]

Many discussions of climate change in popular, and scientific, literature treat a 2°C global mean temperature increase (relative to 1750) as a significant threshold. One reason for this is that it is possible that beyond this threshold our chances of averting various positive feedbacks which will accelerate climate change start to diminish dangerously, and the likelihood of our being able to avoid catastrophic climate change events is reduced.[15] As Mark Lynas puts it,

> If … we cross the 'tipping point' of Amazonian collapse and soil carbon release which lies somewhere above two degrees, then another 250 parts per million of CO_2 will unavoidably pour into the atmosphere, yielding another 1.5°C of warming and taking us straight into the four degree world. Once we arrive there, the accelerated release of carbon and methane from thawing Siberian permafrost will add even more greenhouse gas to the atmosphere, driving yet more warming, and likely pushing us on to the five degree world. At this level of warming … oceanic methane hydrate release becomes a serious possibility, catapulting us into the ultimate mass extinction apocalypse of six degrees.[16]

To have a decent chance of avoiding global warming of 2°C and above requires cuts in global emissions of CO_2 of 50–85 per cent by 2050 so as to achieve stabilisation at 350–400ppm CO_2 (445–490ppm of CO_2-equivalent), with CO_2 emissions peaking between 2000 and 2015.[17] This is the so-called 'peak early/ rapid decline' scenario. There are reasons to be pessimistic with respect to the likelihood that we will stay below the thresholds it contains. Consider the following two facts about the emissions stabilisation targets. First, stabilisation at 350–400ppm CO_2 would deliver warming below 2°C only if climate sensitivity is towards the lower end of the range of likely values,[18] rather than the 'best estimate' of IPCC experts, which actually delivers a range for warming of 2°C–2.4°C. Second, the stabilisation and emissions-reductions targets of the 'peak early/rapid decline' scenario do not take account of positive feedbacks and abrupt climate change created by events such as those identified by Lynas. Indeed, the IPCC lists such 'low probability/high impact events' as a 'key uncertainty' with respect to the 'drivers and projections of future climate changes and their impacts'.[19] And even less dramatic positive feedbacks, created by 'climate-carbon cycle coupling',[20] are not factored in to the emissions reduction target for avoiding 2°C+ warming, which means that this target (and all the others similarly computed) might (in a typically cautious IPCC statement) 'be underestimated'.[21]

If, as seems increasingly likely, we fail to stay below the 2°C target we expose future generations to possible climate change catastrophes as tipping-points are passed, powerful positive feedbacks are unleashed, and average temperature rises above 5°C. However, we need not speculate about climate change catastrophes in order for the writing on the wall for our descendants to come, hazily, into view. Remembering that the IPCC does not take into account 'low probability/high impact' events causing positive feedbacks, or even less dramatic positive feedbacks created by climate-carbon cycle coupling, we can consider the impacts associated with its six stabilisation scenarios for CO_2 emissions,[22] each projecting a range for CO_2 and CO_2-eq ppm concentrations at stabilisation, global emissions of $GtCO_2$ per year en route to stabilisation, and global average temperature increases at equilibria corresponding to the stabilisation levels.[23,24] Using the IPCC's best estimate of climate sensitivity at 3°C, *all* of these scenarios project minimum temperature increases at equilibrium of over 2°C. Using the lowest estimate of climate sensitivity at 2°C, four scenarios project minimum increases at equilibrium over 2°C; and taking the highest estimate of climate sensitivity at 4.5°C, even the 'peak early/rapid decline' scenario – when CO_2-eq emissions stabilise at 445ppm (350ppm CO_2) and peak early – projects a minimum temperature increase at equilibrium of 3°C. Looking at things from the other end, the worst stabilisation scenario projects maximum temperature increases at equilibrium of just under 4°C with the lowest estimate of climate sensitivity, 6.1°C with the best estimate, and over 8°C with the highest estimate.[25]

Perhaps reaching equilibrium will seem a distant event – a set of conditions creating problems for people very different from us, who may have technofixes available, or who will at least have had time to adjust to the myriad changes and difficulties that stabilisation at anything over 535ppm CO_2-eq could bring.[26] So let us consider conditions in which our grandchildren, and/or their children, could live. The IPCC's lowest value in the likely range of temperature increase for the best-case scenario is 1.1°C; the highest value in the likely range of increases for the worst-case scenario is 6.4°C; its best estimates for these scenarios are 1.8°C and 4°C respectively. These projections are for the next 80–90 years.[27] The increases are given relative to 1980–99 levels; for comparison with pre-industrial levels (against which the 2°C threshold is measured by many commentators), 0.5°C should be added to each figure, i.e. 2.3°C and 4.5°C respectively. And remember that these projections do not take account of possible high impact/low probability abrupt events causing positive feedbacks, or those caused by the climate-carbon cycle.

Taking the last category of feedbacks into account, measuring against pre-industrial average temperatures, and working with the latest figures for global emissions increases, consider the UK Met Office's characterisation of the worst-case scenarios for climate change, to which it urges policy makers to turn their attention.[28] On its account, this involves an increase in global average temperature of 7.1°C by 2100 which, it claims, has a 10 per cent chance of occurring.[29] This is an almost unthinkable increase. Virtually no climate models exist for temperature increases at this level, and the speed at which this increase would take place – leaving no time for evolutionary adaptation – is unprecedented in the history of the planet. At 7.1°C we would likely have no chance of adapting to climate change, let alone mitigating it, and would probably be on the verge of extinction, if not already extinct. Indeed, at 6°C climate change would likely be on an accelerated path marked by various tipping-points, and punctuated by abrupt and catastrophic events that would be out of our control entirely. The 6°C+ world really is the worst-case scenario, and it is one in which the possibilities for coordinated action to enable the survival of the species (putting aside any possibility of addressing climate change through coordinated action) would be gone. There is little point in speculating about politics for this kind of world. So, for the purposes of this book, I shall assume a worst-case scenario of temperature increases by the end of this century, and for equilibrium some time in the next few centuries, of 5.5°C, which, according to the Met Office, is the most likely outcome under a business-as-usual scenario (i.e. one in which no action on climate change is taken, and greenhouse gases continue to be emitted at the current rate, with a 50 per cent chance of occurring).[30]

Here are some bare facts about impacts at 5°C and above. Hundreds of millions of people would be exposed to increased water stress; more than 40 per cent of the Earth's species would go extinct; cereal productivity would decrease in low latitudes, causing food shortages (Africa and Asia would lose one third of their food supply); at least 30 per cent of coastal wetlands would be lost and millions of people would be exposed to coastal flooding; risks to human health would multiply, with increased levels of malnutrition, malaria, infectious diseases, and morbidity from heat waves, floods, and droughts.[31] One climate model for a 5°C increase predicts two global 'dry belts' of permanent drought. The northern belt would comprise 'all of the central Americas, the entire southern half of Europe, the western Sahel and Ethiopia, southern India, Indo-China, Korea, Japan and the western Pacific'.[32] The southern belt would encompass 'the southern portions of Chile and Argentina, eastern Africa and Madagascar, almost the whole of Australia and the Pacific islands'.[33] The desertification of these regions would make human habitation there increasingly unfeasible, creating large numbers of environmental refugees seeking admission to shrinking areas of habitable land in northern latitudes. At 5°C the collapse of oceanic methane hydrates becomes, with time, a possibility, as the oceans warm up. Such collapse would release huge amounts of CH_4 (a greenhouse gas twenty times more potent than CO_2) into the atmosphere, and thus increase the likelihood of further temperature increases.

The sea-bed landslides caused by the explosions would also create huge tsunamis with terrible destructive potential.

Putting worst-case possibilities to one side, there are grounds for pessimism that the future we are creating will meet the 'peak early/rapid decline' profile of the best-case scenarios. As earlier noted, CO_2 concentration in 2004 was already at 379ppm, and stabilisation at 350ppm–400ppm is required in the best-case scenario. In the 100-year period 1906–2005, the linear trend for annual global average surface temperature increase was 0.74°C per year, and the rate of warming is speeding up:[34] glaciers are retreating, permafrost is melting, spring is coming earlier, and species are migrating polewards and upwards to find habitats suitable for them.[35] Putting aside the 'peak early/rapid decline' scenario as too optimistic, consider some likely impacts of climate change on human well-being and social organisation over the next 50–100 years under scenarios less frightening than 'business-as-usual'.[36]

The IPCC makes the following regional predictions, each of which it assigns 'high' or 'very high' confidence (i.e. an eight out of ten, and nine out of ten chance respectively of being correct). By 2020 in Africa, between 75 to 250 million people are expected to be exposed to increased water stress, and in some African countries changes to precipitation patterns could reduce food production from rain-fed agriculture by up to 50 per cent, worsening Africa's already terrible problems of malnutrition.[37] By the 2050s in Asia, more than a billion people could be affected by decreased fresh water availability, especially in large river basins. The heavily populated megadelta regions in South, East, and South-East Asia (for example, Bangladesh) will be at increased risk of flooding from the sea, and from rivers, with concomitant increased rates of death from cholera and diarrhoeal disease.[38] In Australia, the Great Barrier Reef will significantly bleach, with a resulting loss of biodiversity, and food production from agriculture will suffer as a result of increased and lengthened droughts, and more frequent fires;[39] similar, although less intensified, heat-related impacts are likely in Southern Europe. In Latin America, by the 2050s the tropical forest in eastern Amazonia will have been replaced by savanna (thus significantly impacting on one of the planet's most important carbon sinks), and many areas of agricultural land will become deserts, with consequences for food security in the region.[40] In North America, forest fires will become larger and more frequent, and heat waves in cities will become more intense.[41] Tropical storms affecting coastal areas are likely to increase.[42] In the Polar Regions, ice and glaciers will thin, and permafrost will melt more extensively and release CH_4; biodiversity in these regions will be badly affected.[43] Perhaps worst affected in the near to medium term will be the small island states. Rising sea levels, more numerous storm surges, and coastal erosion will make habitation in these places increasingly difficult; and widespread coral bleaching, with its effects on fisheries, combined with reduced water resources during more frequent low rainfall periods (in, for example, the Caribbean) will add to these difficulties.[44] Across all these impacts in different regions, the vulnerable and the poor will be worst affected in virtue of being the least able to adapt.

Prospects and policy

Despite the fact that climate change is unequivocal and accelerating, many people still do not see climate change policy as a top priority, and remain unwilling to make even the most straightforward changes to their lifestyles that are required if we are to have any hope of staying below the 2°C target. In a 2007 UK poll, for example, only 19 per cent of people placed the environment as a priority in politics, 6 per cent down on a similar poll conducted in 2001; 25 per cent of those polled agreed with the statements, 'It takes too much effort to do things that are environmentally friendly' and 'I don't believe my behaviour and everyday lifestyle contribute to climate change'; and up to a third agreed that they 'don't really want to' make changes to reduce emissions, such as using their cars less frequently and flying less.[45] And polls in which a high percentage of respondents answer that climate change is an important issue to them nevertheless rank climate change low in their priorities compared to other issues, such as terrorism: for example, a Pew Centre poll showed that as many as 75–80 per cent of US respondents assigned importance to climate change, yet placed it twentieth out of twenty, compared to other issues.[46]

There are many sociological and psychological explanations as to why what public concern there is about climate change does not manifest into action that tracks the seriousness of the phenomenon. Human beings' assessments of risk reflect features of their situation often not related to the objective severity and nature of the risk itself.[47] Furthermore, and perhaps worse, the hegemony of consumerism in developed nations (and its nascent appearance in developing nations such as China),[48] which do most to contribute to greenhouse gas emissions, combined with the rapidity with which hitherto energy-intensive luxury goods are being seen as necessity goods (for example, in 1996, 32 per cent of people in the US thought that a microwave was a necessity, whereas in 2006, 68 per cent believed this),[49] creates a culture in which greenhouse gas reductions are difficult to achieve, and simpler living is seen as the lifestyle choice of eccentrics.[50]

In the face of such obstacles decisive, immediate, coordinated and effective political action on climate change is required at all levels, from the global through the national to the local. In the early 1990s there were some grounds for optimism about the prospects for such action. In 1990 the IPCC published its First Assessment Report, presenting the scientific evidence for the existence of anthropogenic climate change, and grounding calls for political action on climate change; and two years later the United Nations Framework Convention on Climate Change (UNFCCC, or the so-called 'Rio Convention') was adopted (by 192 nations) at the 1992 Rio Earth Summit (it came into effect in 1994). The UNFCCC is a framework document setting the rules and procedure for further action. As such, it sets no targets or deadlines for parties (developed nations were given the option of adopting the voluntary goals of reducing their emissions to 1990 levels), but rather, states their joint recognition that anthropogenic climate change exists and requires political action, and their shared commitment to principles of precaution, sustainability, equity, and cooperation for the benefit of current and future generations in attempts to address the problem. In particular, the UNFCCC states a commitment to 'common but differentiated responsibilities' in tackling climate change, which is supposed to ensure fairness in the distribution of the national emissions reductions needed in order to achieve the necessary global reductions. The UNFCCC is bursting with ideals and statements of good intentions. The negotiations surrounding the Kyoto Protocol – adopted at the third conference of the parties (COP) to the Rio Convention in 1997, and eventually entering into force in 2005 – brought those enthused by these ideals and intentions back down to Earth with a bump.

The Kyoto Protocol sets different legally binding emissions reductions targets for parties to the UNFCCC. So-called 'Annex B' (industrialised and developed) countries are bound to reductions which would amount to an overall 5.2 per cent reduction (against 1990 levels) in greenhouse gases by 2012. At the insistence of the USA, the Kyoto Protocol also included 'flexibility' mechanisms enabling Annex B countries to meet their targets through means other than domestic reductions: for example, the international emissions trading regime enables countries to buy and sell emissions permits, and the Clean Development Mechanism enables Annex B countries to get emissions reductions credit for investing in emissions-reduction projects in developing nations. The COP-3 negotiations did not deliver any mandatory targets for developing nations, and US political resistance to the Protocol in virtue of this feature – on the grounds of its unfairness to the USA, despite the fact that the US was, and remains, the world's biggest per capita emitter – became more entrenched and vociferous over the ensuing negotiations at COP-4 and COP-5. Conflict between the US and the other parties – in particular, the EU – came to a head at COP-6 in The Hague in 2000 as the US insisted on higher emissions allowances for trading, and on greater credit against its reductions targets for its carbon sinks (in its large areas of forest). In 2001 the newly elected US President George W. Bush withdrew the US from the Protocol. Although, with 25 per cent of the world's emissions, US withdrawal was not sufficient to kill the Protocol – which required (and got) ratification by 55 countries representing 55 per cent of global emissions – it was sufficient to make the Protocol toothless. Subsequent amendments to the Protocol – increasing emissions permits, weakening penalties for noncompliance, and increasing credit for carbon sinks – ensured sufficient ongoing participation by countries for the ratification threshold to be met in 2004, when Russia ratified the treaty. On any account of the matter, the saga of the Kyoto Protocol is a depressing tale, and the emissions reductions targets it set – even absent the various 'flexibility mechanisms' – were totally inadequate to address the problem recognised by the UNFCCC.[51] Perhaps most depressing of all is the amount of time that has been spent thrashing out the details of this inadequate treaty; each successive set of IPCC Assessment Reports shows the climate change problem to be worse than we thought, and the amount of time left to do something about it to be less and less.

At the most recent conference of the parties – COP-15 in Copenhagen – no agreement was reached on whether to adopt a successor to the Kyoto Protocol, and no further legally binding emissions reductions targets were set. At this meeting the parties 'took note of' (but did not adopt) the so-called 'Copenhagen Accord', which does no more than reiterate the ideals of the Rio Convention and the inadequate commitments of the Kyoto Protocol. Parties were subsequently invited (by the UNFCCC Secretariat) to pledge emissions reduction targets or actions (for non-industrialised countries), relative to a baseline of their choosing, with a 'soft deadline' of 31 January 2010. The COP-15 meeting was almost universally hailed as a failure. The consequences of this for, for example, the small island state of the Maldives were made clear by its president, Mohamed Nasheed, in a speech to the summit on climate change at the UN headquarters in November 2009, just prior to the Copenhagen meeting.

> Please ladies and gentlemen, we did not do any of these things, but if things go business-as-usual we will not live, we will die. Our country will not exist. We cannot come out from Copenhagen as failures. We cannot make Copenhagen a pact for suicide. We have to succeed. We have to make a deal in Copenhagen.[52]

The current generation's track record with respect to climate change policy is dismal and shameful. However, this is not a book about climate change policy. The arguments it contains are pitched at a higher level of abstraction, and I do not use them in analysis and critique of policy approaches.[53] Keeping in mind the policy failures to date, the increasing seriousness of the problem as time passes and opportunities to take effective action disappear for good, and the worst-case scenarios with which we flirt as our representatives bicker and prevaricate at one COP meeting after another, gives increased impetus to radical principles of justice fit to address the climate change problem, such as those I defend here.

Plan of the book

Let me briefly indicate how the book proceeds, and how the arguments it contains are related to one another.

In Chapter 2 I lay out the theoretical apparatus made use of throughout by giving an account of John Rawls' approach to intergenerational justice, which draws on commitment to a particular form of political argument. I give consideration to how this ideal can guide the deliberations of experts involved in climate change policy making, and consider some worries about the role of experts so conceived. This chapter also includes a response to the notorious 'non-identity' problem which plagues most accounts of intergenerational justice, and it concludes with an affirmation of the importance of ideal theory in political thinking.

In Chapter 3 I offer two Rawlsian arguments for a strong version of the precautionary principle fit to inform policies aiming at the mitigation of climate change. These arguments draw on reflections about what any current generation owes to future generations, first, as a requirement of intergenerational distributive justice, and second – and more fundamentally – as a requirement of a commitment to the possibility of justice for all generations. The core of this argument is that no generation is permitted to act so as to jeopardise the possibility of justice for generations that follow. Because the current generation's emissions do just this, they are impermissible in the name of the precautionary principle.

Chapter 4 addresses the question of what justice requires if – as is likely – the current generation fails to mitigate climate change as demanded by the precautionary principle. The argument here draws on principles of corrective justice applied to the intergenerational context, and defended as appropriate to this context using the resources of Rawlsian political liberalism. The upshot of the arguments made in this chapter is that any generation imposing impermissible risks of serious climate change harm on future generations is liable for harm to future generations that might mature from these risks, and that the best way to secure corrective justice for future generations thus put at risk is to require any risk-imposing generation to contribute to an Intergenerational Climate Change Compensation Fund to which members of future generations may apply if and when they suffer climate change harm. In the course of making these arguments I elaborate on the sense in which any generation is liable for the imposition of impermissible risks, and consider what justifications and/or excuses are available to such a generation.

Chapter 5 addresses what might happen to politics if the worst-case climate change scenarios come to pass for future generations. In the nature of the case, the arguments made here are more inchoate and tentative than in the preceding chapters. I consider the nature and justification of principles of triage that might be appropriate in worst-case contexts, and whether such principles are principles of justice. I argue that, although not principles of justice in themselves, principles of triage ought to be at least in part justified in terms of the extent to which they might keep alive hope for a future return to justice for people following the unfortunate generation(s) forced to practise triage.

Chapter 6 concludes with reflections on how to motivate people to do climate change justice to future generations, from within a Rawlsian framework. Here, I draw on how membership of a social union of social unions aiming at the achievement of justice across generations provides members of each generation with social bases of self-respect. Given this, I argue that the current generation's failure (so far) to address the climate change problem so as to contribute to the intergenerational pursuit of justice provides grounds for shame: it is no exaggeration to say that this failure would mean that the current generation fails to do justice to the whole human race.

Before getting started, let me note that I am conscious that much discussion in the book focuses on worst-case scenarios, and that climate change catastrophes play a prominent part in many of the arguments made. This gives the book a rather bleak, if not apocalyptic, feel. However, that is the nature of the beast: there are no good-news stories with respect to climate change (at least, not yet). The possible worst-case scenarios of climate change have prominence because a Rawlsian orientation directs us to attend to the position and circumstances of the worst-off in various sets of outcomes associated with different principles of justice. In intergenerational terms, the worst-off in the worst outcome of the worst-case (business-as-usual) scenario for climate change are very badly off indeed. As already indicated, climate change catastrophes are not science fiction, and policy makers ought to be taking them far more seriously than they do at present, especially if they are committed – as they ought to be – to intergenerational justice. Many of the arguments in this book present principles of justice that ought to inform such serious consideration.

2

A RAWLSIAN APPROACH TO CLIMATE CHANGE JUSTICE

Introduction

In this chapter I lay out the key features of the approach in normative political theory that I shall use throughout the book in analysis of the problems of justice posed by various aspects of climate change, and in framing suggestions for solutions to them. I shall not defend this approach against its competitors, nor shall I unpick all the detailed technical debates spawned by each of its elements. Rather, I shall indicate the political values, principles and ideals central to the approach, and the way in which these values, principles and ideals can impact on thinking about the political problems of climate change in a novel and productive way. Readers with no interest in political theory at its most abstract might prefer to skip to subsequent chapters, although the full defence and elaboration of arguments made in these chapters will require familiarity with the material in this one.

The normative approach I shall use throughout this book in making arguments is political liberalism, which I conceive of as composed of two key elements: a method of political justification, and a commitment to political principles of freedom and equality, with intergenerational scope. It is these aspects of political liberalism (or 'liberalism', as I shall often refer to it) that I shall mine to provide insights into the requirements of justice with respect to the political problems of climate change. However, liberalism does – of course – have implications for, and indeed explicit positions on, a variety of political questions beyond those of justification and principles. For example, there are debates within liberalism about the agents of justice (are they individuals? states? peoples? nongovernmental organisations or multinational corporations? all of these?), the best vehicles for achieving justice (regulation? structured incentives within markets? statute? all of these, in some combination?), and the territorial scope of justice (limited by national boundaries? transnational? global?). Here, and throughout, I shall give no systematic account of answers to any of these questions (although I shall, as the need arises, discuss them in an ad hoc manner).

The liberalism to which I shall appeal is heavily informed and influenced by John Rawls. However, the theoretical approach adopted throughout is not doctrinaire. Where elements of Rawls' approach to an issue are not illuminating they will not be adopted; and elements that are illuminating will sometimes be adapted. And on issues about which Rawls remains silent, I shall sometimes invoke values and ideals which are consistent with Rawls' conception of justice, but certainly are not part of the 'party line'. Thus, I make no claim that all the arguments made in this book are an interpretation of Rawls; what matters instead is whether they stand as arguments. If they do, I would place them in the Rawlsian tradition, but in the end nothing of much importance hangs on this.

The next section lays out Rawls' approach to political justification in terms of public reason, and indicates ambiguities in the account which make it possible to apply this model of political justification to questions of climate change policy. The section 'Experts in democratic politics' makes suggestions for qualifications to this model of public reason enabling it to accommodate and protect the essential role of experts in political debate about climate change policy; this section also articulates and assuages some worries about democratic politics created by these suggestions. The section 'Intergenerational justice' gives an account of Rawls' principle of intergenerational equality, the just savings Principle, and the subsequent section 'Interlude: the non-identity problem' addresses a notorious problem which, according to some, stymies attempts to theorise intergenerational justice. The section 'Ideal theory, background justice, and the circumstances of justice' provides comment on the nature of Rawls' theory as 'ideal', and indicates how such theory can work even in the terribly nonideal circumstances which have created and which exacerbate climate change.

Political justification

Rawls' ideal of democratic political justification consists of arguments for principles of justice which are acceptable, at least in principle, to as many people as possible in conditions of reasonable pluralism (so as to make political justifications inclusive), but which are also fit to support criticism of the political status quo, if required (so as to make justifications critical). A political justification realises this ideal, *inter alia*, when it is such that no reasonable person could reject it.

Rawls claims that the reasonable person is reasonable in two key ways: politically and philosophically.[1] Politically reasonable people exercise their capacity for a sense of justice so as 'to understand, to apply, and to act from the public conception of justice which characterises the fair terms of social co-operation'.[2] When people exercise this capacity they seek publicly acceptable and mutually advantageous solutions to shared problems of social cooperation: they are not stubborn, manipulative, dishonest, or perverse. And the fairness of terms of social cooperation they propose is to be judged according to how far they protect the (political) freedom and equality of citizens. Furthermore, politically reasonable people uphold the 'liberal principle of legitimacy' whereby each of them strives 'to live politically with others in the light of reasons all might reasonably be expected to endorse'.[3] A commitment to such public reason reflects the fact that 'public justification is not simply valid reasoning, but argument addressed to others'.[4] This means that when we engage in political justification we must '[proceed] correctly from premises we accept and think others could reasonably accept to conclusions we think they could also reasonably accept',[5] always bearing in mind that other citizens have the same right to political freedom and equality as we do. However, until we know the conditions in which people committed to the use of public reason converse, it is unclear how public reason secures critique; that is, it is unclear what restrictions commitment to public reason places on the considerations that can be exchanged in democratic political justification unless we know the nature of the disagreements people have that make politics necessary.

Rawls thinks that in any democratic society in which basic rights and liberties are protected people will use their reason freely to form different views on questions of religion, morality, philosophy, and politics: 'the diversity of religious, philosophical, and moral doctrines found in modern democratic societies is not a mere historical condition that may soon pass away; it is a permanent feature of the public culture of democracy'.[6] The explanation for this is that free reason stands under certain 'burdens of judgement' which make disagreement between persons on religious, moral, and philosophical questions inevitable, expectable, not to be regretted, and reasonable.[7] These are that:

a. 'The evidence – empirical and scientific – bearing on the case is conflicting and complex, and thus hard to assess and evaluate'.
b. Agreement on the considerations relevant to a particular judgement does not guarantee agreement with respect to how these considerations should be weighted in judgement.
c. The concepts appearing in judgement 'are vague and subject to hard cases', making interpretation unavoidable.
d. 'To some extent … the way we assess evidence and weigh moral and political values is shaped by our total experience, our whole course of life up till now: and our total experiences must always differ.'
e. Normative considerations in cases of judgement are hard to assess, both on their own, and in relation to one another when they are in competition.[8]

According to Rawls, a philosophically reasonable person accepts the existence of the burdens of judgement, which means accepting that – with the best will in the world, and without epistemic failure on anyone's part – agreement between herself and reasonable others on the big (and for that matter, small) questions is not inevitable, and indeed may never be achieved.[9] Philosophically reasonable persons also accept the consequences of the burdens of judgement for the exercise of public reason, he thinks, by limiting the considerations they offer to one another in justification of fundamental principles and/or constitutional essentials to 'common sense', 'plain truths', and 'the methods and conclusions of science when not controversial'.[10] The permanence of reasonable pluralism means that public reasons must '[stay] on the surface, philosophically speaking':[11] public reasoning must remain epistemically abstinent with respect to moral, religious, and philosophical matters about which reasonable people using their reason freely may disagree.[12] Justificatory debate may appeal to 'the general beliefs and forms of reasoning found in common sense, and the methods and conclusions of science, when not controversial' but

> [T]his excludes comprehensive religious and philosophical doctrines (the whole truth, as it were) from being specified as public reasons … [I]f we are to speak of public reason, the knowledge and ways of reasoning – the plain truths now common and available to citizens generally – that ground … the principles of justice must be accessible to citizens' common reason.[13]

Political liberalism's epistemic abstinence on the big questions ensures that no person must repudiate the moral, religious, or philosophical outlook that shapes her life in order to subscribe to principles of justice (which makes its political justifications inclusive). And the requirement – argued by political liberals to be consistent with epistemic abstinence – that principles must be justifiable in public reason secures for justification the desired critical edge. Public reason always stands as a possible corrective to political justifications which do no more than express the values of the status quo or, worse, which deliver principles that only serve the interests of political and/or economic elites.

In sum, this ideal of political justification requires that people accept that others will reasonably disagree with them on questions of religion, philosophy, value, and other matters relating to their comprehensive conception of the good and worldview that are not accessible to 'citizens' common reason', and that people signal this acceptance through a commitment to propose and abide by principles of justice justified in public reason, so long as others do so too, and thereby signal their recognition of their own and fellow citizens' freedom and equality. This is a deeply democratic vision of the terms of shared political life. Because no person is required to become a 'true believer' of any sort in order to engage in the practice of political justification, and because each person is required to recognise the rights that others have to so engage insofar as they are reasonable – even if all others disagree with her – every person is a potential member of the constituency of justification, and thus all are equal in their opportunities to influence the political decisions of their community.[14]

Let us pause here to note an implication of this democratic ideal for thinking about climate change. It could be – indeed, it has been – argued that adequate political solutions to the problems of climate change must draw on an ecological theory of the good, and an accompanying vision of social and political life radically reconfigured from that which currently exists in late capitalist societies.[15] 'Ecology' in political thinking is a contested term, but when it has social and political implications it almost always includes an 'ecocentric' theory of value, whereby nonhuman life (and sometimes also inanimate objects, and systems involving inanimate and animate elements) has value independent of any human valuations of it. If ecocentrism is a theory of value about which it is reasonable to disagree (and I think that this is the case), then fundamental political principles cannot be justified by reference to ecocentric values, on the Rawlsian vision of democratic politics just described. It is not my purpose to elaborate or assess ecocentrism.[16] Rather, my point is that, despite the ambitious and inventive blueprints for social and political change that often accompany affirmation of ecocentrism – indeed, despite the promise of many of them to tackle problems of climate change at their roots – an ecocentric theory of value is ruled out by a commitment to Rawlsian liberal democracy. Importantly, this is *not* to say political justification according to the Rawlsian ideal implies that ecocentrism is false and that anthropocentrism is true. Rather, it is to say that political justifications guided by this ideal are – and ought to be – silent on this philosophical question, because disagreement with respect to it is reasonable.

Why is it important to uphold the Rawlsian ideal? A long discussion of the moral significance of the inclusivity aimed at by this model of 'political' liberal justification would take us too far from our subject. However (and I shall draw on these reflections later in the book), it is worth noting the support that would be given to the self-respect of persons engaged in the practice of political justification conducted on political liberal terms, along two dimensions. For Rawls,

> [Self-respect] includes a person's sense of his own value, his secure conviction that his conception of the good, his plan of life, is worth carrying out. And second, self-respect implies a confidence in one's abilities, so far as it is within one's power, to fulfil one's intentions. When we feel that our plans are of little value, we cannot pursue them with pleasure or take delight in their execution. Nor plagued by failure and self-doubt can we continue in our endeavours ... Without [self-respect] nothing may seem worth doing, or if some things have value for us, we lack the will to strive for them. All desire and activity becomes empty and vain, and we sink into apathy and cynicism.[17]

The significance of self-respect not only for persons' emotional and psychological well-being, but also for the exercise of their practical reason, explains Rawls' claims that 'self-respect is perhaps the most important primary good',[18] and that 'parties in the original position would wish to avoid at almost any cost the social conditions that undermine self-respect'.[19,20] Rawls calls social conditions fit to support the development of self-respect the 'social bases of self-respect', and throughout his work they are included on the list of primary goods providing the currency of justice.

> [S]elf-respect depends upon and is encouraged by certain public features of basic social institutions, how they work together and how people who accept these arrangements are expected to (and normally do) regard and treat one another. These features of basic institutions and publicly accepted (and normally honoured) ways of conduct are the social bases of self-respect.[21]

There are two sites in which the social bases of self-respect are found: in the framework of the institutions of the basic structure, and in associational life in the diverse communities of shared interests that flourish in the interstices between the institutions of the basic structure. And there are two dimensions according to which these sites can be assessed as social bases of self-respect: the extent to which they provide access to, and the patterns of distribution of, tangible primary goods necessary for the development of self-respect; and the extent to which they enable and encourage mutual recognition between persons necessary for the development of self-respect.[22]

Political justification conducted in terms of public reason, as outlined above, stands as a recognitional basis of self-respect found in the framework of the basic structure. Political society is, as Rawls sometimes puts it, a 'social union of social unions':[23] a cooperative endeavour to achieve justice so as to promote the good of every participant. An important aspect of this good is self-respect, and this can be supported by activity in the private associations of society wherein persons pursue their particular interests and interact with one another so as to exhibit qualities and skills fit to generate mutual respect and esteem.[24] However, it can also be supported in political society, wherein people come together as citizens in pursuit of justice rather than as private persons acting in accordance with their individual conceptions of the good. Here, mutual respect is evinced in inclusive public justificatory debate about matters of justice, couched in the language of public reason. The reason why such political justification supports self-respect is that this depends on a person's conception of herself as worthy of being given justifying reasons by any other person capable of having expectations of her that she act in a certain way. A self-respecting person will demand – or at least believe herself entitled to demand – a justification for another's (explicitly stated, or implicitly assumed) expectation that she do this or that; it is in this sense that self-respect involves non-subservience.[25] Political society is an association of people with just such expectations of one another, generated by a shared commitment to principles of justice. If they are to find the bases of self-respect in their association, then they must at least stand ready to (if not actually) provide one another with justificatory reasons for these expectations that each will act as justice requires. And if citizens are also required to take seriously the burdens of judgement, then these reasons cannot be framed just in terms of the comprehensive conceptions of the good which differentiate them, and must meet the standards of public reason. When a person is excluded from such debate – for example, by having her reasons ignored, by not being offered reasons at all, or by being offered reasons by others who are not sincere in their attempt to practise public reason – then her status as a person worthy of giving and receiving justificatory reasons is undermined, and the non-subservience partly constitutive of her self-respect comes under threat.[26]

It is often asserted that Rawls' account of public reason as the required language of political justification relates only to debate about fundamental principles of justice and constitutional essentials, and not to debate about legislation and policy. If this is true, then reflection on these principles may have little to tell us about what justice requires with respect to the problems of climate change. This common assertion does not attend to some ambiguities and vagueness in Rawls' discussion of the forums, scope, and subject matter of public reason which would permit – indeed, might require – that it be extended to address (some) debates about policy. By noting these ambiguities and their implications a case can be made for judging political debate and decision making on climate change according to the standards and values of public reason, and thus the scene is set for the arguments to follow in subsequent chapters that do just this.

In Rawls' vision, political justification occurs in three arenas: fundamental principles of justice, the constitution, and statute. Justificatory debate in each arena is constrained by the outcome of debate in the tier above it; for example, any legislative decision regarding the use of nuclear weapons is bound not to violate the rights guaranteed to citizens by a constitution which encodes the fully adequate scheme of basic liberties laid out by the first principle of justice.[27] Rawls claims that his ideal of democratic political justification in terms of public reason

> [Applies] to questions concerning constitutional essentials and matters of basic justice … to citizens when they engage in political advocacy in the public forum, in political campaigns for example and when they vote on those fundamental questions … to public and government officers in official forums, in their debates and votes on the floor of the legislature [and] especially to the judiciary in its decisions and as the one institutional exemplar of public reason.[28]

Thus, it would appear that public reason does not apply to common or garden political debates that are the bread and butter of ordinary politics. To take Rawls' own examples, 'much tax legislation and many laws regulating property; statutes protecting the environment and controlling pollution; establishing national parks and preserving wilderness areas and animal and plant species; and laying aside funds for museums and the arts'.[29] It is consistent with a range of political societies all being just (insofar as they satisfy the principles of justice in their constitutions and statutes, and achieve this through democratic debate in public reason) that they make different legislative decisions on these matters: equally just societies could be markedly distinct in terms of how they answer non-fundamental political questions at the sub-constitutional level, and could feel that way to the citizens in them.

However, Rawls notes that he has not given a full account of the nature and significance of the differences between legislative debates related to the issues listed above, and debates about principles of justice, the constitution, and statute as it relates to these things, such that the application of public reason to the latter subjects but not the former is well grounded. The partial account that is given relates to the fact that the latter involve questions of basic or fundamental justice (i.e. justice as it is related to the first principles of justice) whereas (most) debates about statute and policy do not. However, in the same breath as excluding legislative and policy debate from the scope of public reason he allows that 'sometimes these [questions] do involve fundamental matters'.[30] And immediately after this discussion he goes on to explain his focus on principles and the constitution (and statutes insofar as they realise and uphold these things) in terms of an aim to

> [C]onsider first the strongest case where the political questions concern the most fundamental matters. If we should not honor the limits of public reason here, it would seem we need not honor them anywhere. Should they hold here, we can then proceed to other cases.[31]

Thus, it is not clear that all legislative decisions would always be excluded from the scope of public reason, given a full account of the matter.[32] Some may directly involve matters of fundamental justice, or turn out to do so upon a close inspection of their consequences for other areas of legislation; some statutes in a certain area may have this character while others in the same area do not. And there may be a case for the application of public reason to even those legislative debates which we can say for certain raise no, or have no implications for, matters of fundamental justice. The form of political justification that most of us encounter directly is legislative debate about non-fundamental matters; very few of us have contact with constitutional conventions or arguments about amendments, and even fewer participate in rarefied discussions about the justification of fundamental principles of justice in political philosophy. Adopting a commitment to public reasoning about legislative matters could deliver to people a tangible experience – on the ground, as it were – of the values of equality and reciprocity embedded in the ideal of democratic political justification to which politics in higher tiers always (ought to) answer. If we care (as Rawls does, very much)[33] about the extent to which a democratic society can generate allegiance to it among its members *sui generis* – just through their experience of living together on just terms in such a society – then it is worth attempting to ensure that they can see public reason in action in at least some legislative debates.

My interest here is in the politics of climate change, which usually manifests in legislative debate about measures to protect the environment. There are two ways in which to justify the use of public reason to address the questions these debates involve. First, and indirectly, legislation to tackle climate change can raise fundamental matters of justice, in particular, issues related to intergenerational justice (as laid out in the section 'Intergenerational justice'). When this is the case, appeal to the relevant principles of fundamental justice, and discussion of their consequences in public reason, can yield specific recommendations to adopt one sort of policy over another: for example, in Chapter 4 I argue that Rawls' just savings principle has an important role to play in an argument for building up a fund to provide compensation, in the name of corrective justice, to future generations harmed by climate change. The second, more direct, way in which to justify the use of public reason to argue about climate change does not appeal to matters of fundamental justice affected by climate change policy, but rather to the fact that climate change beyond a certain point destroys the possibility of fundamental justice altogether; here, the appeal is to the specific effects and unique character of climate change itself. For example, in Chapter 3 I argue that precautionary measures to mitigate climate change are justified because failing to adopt these measures imposes a risk on future generations of the destruction of the possibility of justice, and this cannot be justified in public reason (similar reflections also play a role in Chapter 5, which discusses triage). In the section 'Ideal theory, background justice, and the circumstances of justice' I shall lay out an account of climate change and the circumstances of justice on which these subsequent arguments will draw.

For now, let me turn to a different problem. Despite the fact that public reason could, and perhaps should, be extended to cover debates about climate change legislation, we might nevertheless think this is not possible. The epistemic abstinence built into the ideal of democratic justification excludes from political debate scientific (and other expert) judgements which are essential to the formulation of such policy because these judgements are not a product of 'the general beliefs and forms of reasoning found in common sense'[34] to which debate in public reason must be restricted. Because scientific (and other) expert opinion does not consist of 'plain truths' – and because the methods and conclusions of scientific (and other) experts are never uncontroversial[35] – policy makers using Rawls' public reason prima facie would not get very far. There is, it seems, an unavoidable mismatch between the essentially 'nonpublic' reasons of communities of experts which must inform debate about climate change policy, and the essentially public reason in which, I have claimed, such debates must be couched (at least sometimes). The next section considers steps to address this problem, and assuages some worries.

Experts in democratic politics[36]

Rawls claims that 'public reason is the form of reasoning appropriate to equal citizens who as a corporate body impose rules on one another backed by sanctions of state power', and contrasts it with nonpublic reason, which is 'appropriate to individuals and associations within society; it guides how they quite properly deliberate in making their personal and associational decisions'.[37] Public reason supplies the language of the highest order of justificatory debate in a socio-political context, whereas nonpublic reason is appropriate in the smaller voluntary associations of civil society. Rawls' sharp division between public and nonpublic reason supports an analogy that he draws implicitly between devotees and scientists (and, by extension, other experts). Although he acknowledges that '[t]he nonpublic reasons of churches and universities, of scientific associations and private clubs differ',[38] he repeatedly mentions scientists qua members of associations in the same breath as religious believers qua members of churches: their associational reasons and reasonings, qua nonpublic, ultimately have no place in socio-political justificatory debate governed by public reason.[39]

Prima facie, this restriction on public reasoning means that it cannot be used in policy making and legislative debate: there is no way to justify draft legislation without reliance on the conclusions of experts reasoning together in thoroughly nonpublic and highly specialised, often inaccessible, ways.[40] As such a policy matter, the question of how to tackle climate change is not like the question of whether to permit gay marriage, or whether to ban Muslim headscarves in state schools:[41] it essentially involves consideration of more than just competing values expressed in fundamental principles of justice, and requires detailed understanding of the processes and likely effects of climate change in a huge array of areas, from meteorology to marine biology, from epidemiology to atmospheric aerosol science. This is why organisations such as the IPCC – which was established by the United Nations (UN) and World Meteorological Organization in 1988, and reports on the causes, impacts, and options for tackling climate change every five years or so – were set up. Ignoring the nonpublic reasoning of such scientific associations would leave policy makers completely in the dark with respect to framing legislation to tackle climate change. It cannot be that, just as the pronouncements of priests should be kept out of public debate about gay marriage, so the (often) controversial research of atmospheric scientists ought to remain firmly within the limits of university departments of meteorology.

There are two ways forward. First, we might simply deny that public reason has any application to legislative matters related to climate change and instead accept the rough and tumble of interest-group politics as the best route to legislation, so long as it is reined in by a constitution that realises the democratic ideal. However, as indicated in the last section, I want to consider extending the ideal of public reason to policy making that makes ineliminable reference to the nonpublic reasoning of communities of experts. An important part of this departure involves rejecting the analogy between scientists and devotees. With respect to climate change policy, the advice of the IPCC is not like rabbinical counsel: it is not illegitimate for policy makers to appeal to it, and it is anyway not dispensable in making climate change policy. But neither is scientific advice on climate change akin to 'the general beliefs and forms of reasoning found in common sense',[42] and it is certainly not uncontroversial, even among scientists themselves. For example, take the phenomenon of global dimming, whereby the amount of sunlight reaching the Earth's surface declined by 5 per cent between 1960 and 1990.[43] The cause of global dimming is an increase in aerosol particles – pollution – in the atmosphere. Anti-pollution measures are reversing global dimming, but this increases global warming. Common sense tells us that anti-pollution measures are good for the environment; science tells us that this is not always the case, all things considered.[44] If climate change science is (properly) indispensable to making climate change policy, and we aim to give an account of public reason fit to govern justificatory debate about (some) legislative matters, then the fact that climate change science does not express 'plain truths ... accessible to citizens' common reason'[45] means that the ideal of public reason must be made more permissive with respect to the considerations it is legitimate to invoke in political justification directed towards legislative questions.

I suggest that the account of public reason so far articulated should be qualified so as to permit the introduction of considerations that are not 'plain truths', and reasons that are not a product of 'common sense', and/or 'the methods and conclusions of science when not controversial'.[46] Public reason should permit the introduction of complex, specialist, often inaccessible, nearly always controversial, and thoroughly nonpublicly reasoned expert advice through the following qualifications:

- *The Permissibility of Expertise*: When empirical and scientific evidence is indispensable to the case, expert judgement is required and – *ipso facto* – acceptable in public reasoning;
- *The Autonomy of Experts*: Although expert judgement is subject to burdens, experts themselves are the best judges of the limits of reasonable disagreement in their area of expertise.

The Permissibility of Expertise requires that legislators working within the limits of public reason take counsel from experts where appropriate; and *The Autonomy of Experts* ensures a degree of autonomy and independence for experts, which is indispensable to their authority qua experts.

The Permissibility of Expertise and *The Autonomy of Experts* refer to 'expert judgement'. Policy makers guided by *The Permissibility of Expertise* and *The Autonomy of Experts* by definition are not experts, and so the judgements they consult cannot take the form of unadulterated scientific opinions, or unmediated presentation of the results of highly specialised experiments; we can expect policy makers to acquire an understanding of the basic vocabulary and methods in an area of expertise, but to require them to become experts themselves defeats the purpose of the qualifications, and is inefficient. Experts must interpret and present theories, predictions, analyses, and the data underlying them so as to make them reasonably accessible to serious and well-informed policy makers. There are at least three key areas in which experts exercise these skills:[47]

a. Experts present a construction of the problem in hand by specifying the potential damage constitutive of it, and the significance of this damage; or they validate an existing construction of the problem by supplying data fit to support it (call these 'damage-and-data judgements').
b. Experts lay out the technical options for tackling the problem, and their feasibility (call these 'options-judgements').
c. Experts prepare and present data in a way fit to support the assignment of responsibilities for the problem through political processes (call these 'responsibility-judgements').

To see the difference *The Permissibility of Expertise* and *The Autonomy of Experts* make to thinking about policy in public reason, consider the following case. Climate change deniers – i.e. those who deny that climate change is occurring, or claim that its rate and effects have been greatly over-exaggerated – very often cite a 1992 study by John Christy of the University of Alabama in Huntsville that putatively shows that the lower troposphere has cooled since satellite records began in 1979. This supports climate change denial because it creates uncertainty about the accuracy of data showing that the middle and upper tropospheres have warmed since 1979 and about its status as evidence for climate change: how could the lower troposphere be cooling while the middle and upper tropospheres warm up? Does Christy's analysis provide grounds for doubting the reality of climate change? This question must be answered in order for policy makers to consider whether climate change is a problem in need of a policy (let alone for consideration of what that policy should be) and activates *The Permissibility of Expertise.* Three recent independent studies have provided the answers that policy makers directed by *The Permissibility of Expertise* need. The satellite data used by the Christy study had not been corrected for 'satellite drift' – whereby measurements taken at the equator shifted over time, from 2pm to the cooler time of 5pm, and so delivered lower temperature readings; the data from weather balloon temperature sensors used in the Christy study had not been corrected for the installation of improved shields to protect sensors from direct sunlight, which meant they delivered lower temperature readings over time; and nineteen climate models using different physics all predict warming in the lower troposphere.[48] In virtue of the damage-and-data judgements of experts conflicting with it, the Christy study provides no grounds for denial of climate change.[49] With climate change established as a problem through such damage-and-data judgements, the explicit purpose of (for example) the IPCC is to coordinate experts so as to provide policy makers with options- and responsibility-judgements:

> [T]he IPCC embodies a unique opportunity to provide rigorous and balanced scientific information to decision makers. By endorsing the IPCC reports, governments acknowledge the authority of their scientific content. The work of the organization is therefore policy-relevant and yet policy-neutral, never policy-prescriptive.[50]

The activation of *The Permissibility of Expertise* requires policy makers to consult experts.[51] This granted, climate change deniers are likely to counter that it is illegitimate to restrict the set of experts to (what we might call) climate change believers. Why shouldn't legislators lend their ears to climate change deniers as well? There is nothing stated in the qualifications that specifically precludes this. And there is nothing in the ideal of public reason that rules it out either. Indeed, climate change deniers might complain that legislators operating within the confines of this ideal who listen only to bodies of experts such as the IPCC violate the ideal itself by illegitimately privileging one set of comprehensive judgements over others with a more negative tone.

The justification for listening to the IPCC and ignoring climate change deniers should be given by reference to *The Autonomy of Experts*, for the following reasons. Climate change scientists are overwhelmingly in agreement on the following two propositions: (1) that the global average temperature of the twentieth and twenty-first centuries is warming; (2) that the causes of twentieth and twenty-first century warming climate change are anthropogenic.

With respect to (1), analysis of every type of data in the following sets indicates a warming trend in the twentieth century:

- surface air temperatures (average increase of about 0.74°C over the twentieth century);[52]
- sea-level rise (at an average rate of 1.8 mm/yr since 1961 and at 3.1 mm/yr since 1993);[53]
- sea ice (for example, satellite data since 1978 show that annual average Arctic sea ice extent has shrunk by 2.7 per cent per decade, with decreases in the summer of 7.4 per cent per decade);[54]
- ocean temperatures (an increase of 0.18°C in the top 300mm over the last fifty years);[55]
- glaciers (accelerated retreat in the twentieth century);[56]
- climate proxies (e.g. corals, tree rings, ice cores);[57]
- satellite temperature measurements (a temperature increase of 0.06°C–0.26°C per decade).[58]

With respect to the second proposition on which scientists overwhelmingly agree – the anthropogenic causes of climate change – the natural processes known to affect the climate (for example, orbital variations, tectonic activity, solar activity, and natural/internal climate variability, e.g. El Niño and La Niña) either cannot account for the speed of warming recorded in the twentieth century, or are not recorded as having occurred (for example, a series of massive volcanic eruptions). A survey of the abstracts of 928 articles published between 1993 and 2003 in respectable, peer-reviewed, scientific journals showed that not a single paper denied the existence of anthropogenic climate change.[59] Furthermore, it is only once the effects of human activity on the climate are combined with the effects of these natural processes that a model emerges that corresponds to the observations; the model that includes only the effects of human activity corresponds to the observations far more closely than that which incorporates nothing but natural processes.[60] *The Autonomy of Experts* directs us to take this consensus seriously: denial with respect to (1) and (2) lie outside the limits of reasonable disagreement in climate change science.[61,62]

It will be no news to politicians and their familiars that the advice of experts is indispensable with respect to a whole range of policy arenas. In the UK experts sit on advisory committees such as the Human Genetics Commission and the Agriculture and Environment Biotechnology Commission; they participate in ministerial committees such as the Ministerial Sub-Committee on Biotechnology; they constitute themselves as independent bodies with an explicit mission to guide government policy on various questions involving science, such as the Royal Society; and the direction of their research is very often influenced by their pursuit of awards and grants for work in particular areas offered by the various research councils which are under the statutory control of the Department of Trade and Industry. Policy making as it stands in the UK makes full use of *The Permissibility of Expertise* and *The Autonomy of Experts*. If we were to require that legislators addressing climate change also conform to the ideal of public reason – whether directly or indirectly – then the model of political justification I am advocating for climate change policy would be operationalised.

However, despite the government's best efforts,[63] some suspicion of experts in politics remains, and in the wake of public relations fiascos such as 'climategate', climate change scientists in particular have an image problem. To briefly recount this affair, in November 2009 over 1,000 private emails between academic staff at the University of East Anglia's Climatic Research Unit, and between these academics and leading climate scientists at other institutions, were hacked and leaked. Some of the emails were seized upon by climate deniers as a 'smoking gun' showing that climate scientists had manipulated and/or hidden data in order to support their claims about the existence, speed, and extent of anthropogenic climate change, with particular reference to the notorious 'hockey stick' graph featured in the IPCC's Third Assessment Report, which shows a sharp and exceptional rise in average global temperatures in the second half of the twentieth century.[64,65] Three independent investigations cleared the scientists involved of these charges (although questions were raised about the accessibility of raw data, and the peer review processes that the academics in question were involved in), but mud sticks, and some public doubts remain about the integrity of communities of scientists involved in such research.[66]

Abstracting away from the details of the University of East Anglia's public relations fiasco, there are at least three sets of worries about the influence and effect of experts on democratic politics once *The Permissibility of Expertise* and *The Autonomy of Experts* are accepted.

(i) *The Permissibility of Expertise* and *The Autonomy of Experts* will undermine government 'by the people, for the people, of the people' as expressed in the ideal of public reason.

The thought here is that *The Permissibility of Expertise* and *The Autonomy of Experts* open the door to government by unelected cartels of unaccountable experts whose interests may not coincide with the public good, and whose decisions may not protect the freedom and equality of citizens. In his characterisation of 'administrative rationalism', John Dryzek expresses the worry thus:

> administrative rationalism assumes two complementary kinds of hierarchy. The first subordinates the people to the state. The second puts experts and managers in their properly dominant places in the state's own hierarchy, which is justified on the basis of expertise. The discourse pretty much denies the existence of politics of any sort.[67]

Worries about the democratic credentials of administrative rationalism are legitimate, but *The Permissibility of Expertise* and *The Autonomy of Experts* do not transform the democratic ideal to which they are appended into administrative rationalism. Decision-making power with respect to legislative questions for which empirical and scientific evidence is indispensable remains with policy makers, who are not subordinate to the experts they consult. And policy makers are bound by the liberal principle of legitimacy, which ensures that the people are not subordinate to the state.

However, there are more subtle forms of the worry. First, the placement of experts at the apex of government in administrative rationalism need not be formal or explicit. When expert judgement is assigned a role in reasoning about policy, policy makers themselves may fail in their political duties to interpret risks and potential damages according to social values by postponing any decision on the grounds that all the results are not yet in and that more research into the question is necessary: the procrastination of policy makers can create subordination to experts by default.[68]

In reply, it is worth noting that sometimes it is true that results are not yet in, and that in these cases a 'wait and see' attitude is not necessarily procrastination. And so long as policy makers retain ultimate control over legislative matters, and remain answerable to the liberal principle of legitimacy, there need be nothing anti- or un-democratic about such postponement. Cases must be taken in turn and considered on their merits. It may be, for example, that the British government justifies its vouchsafing of genetically modified organism (GMO) trials by procrastination with respect to evidence that GMOs are unsafe,[69] whereas its cautiously permissive approach to embryonic stem cell research is a genuine reflection of the state of stem cell science.[70]

Perhaps, then, the worry is more sinister: giving experts a role in government – albeit limited to an advisory capacity – can put them at the mercy of industry and, indeed, policy makers whose clandestine pursuit of other interests may be well served by inaccurate claims that the science relating to a putative problem is indeterminate or uncertain.[71]

In its least pernicious form, this process involves the 'capture' of a community of experts by industry and business interests, most commonly through donations to think-tanks constitutive of this community; here, neither industry sectors nor experts may conceive of themselves as standing in this relationship. For example, concerned that 'public policy on environmental issues is increasingly driven by moral crusading, rather than objective science or need', and that 'if we really care about the environment, we will ... not deny "good news" stories',[72] Dr Jennifer Marohasy, Director of the Environment Unit of Australia's Institute of Public Affairs, campaigns for GM foods, dubbing the opposition as 'environmental fundamentalism'.[73] Marohasy has science degrees and describes herself as 'an environmentalist'.[74] There is no reason to doubt her good intentions. However, that one of the funders of the Institute for Public Affairs is the genetically modified foods company Monsanto should raise questions about the organisation's impartiality.[75]

In its most pernicious form the process involves direct state interference with the judgements of experts, and manipulation of the internal structures of their organisations to deliver the desired judgements through, for example, withholding of funds or terminating contracts of employment. For example, allegedly the White House edited a 2002 Environmental Protection Agency report so as to create the illusion of scientific uncertainty about climate change despite statements to the contrary in the original text.[76]

The Autonomy of Experts provides at least the beginnings of a principled way to address the capture of communities of experts, and/or their corruption by state officials. *The Autonomy of Experts* states that experts themselves are the best judges of the limits of reasonable disagreement in their area of expertise: with respect to the GM foods debate, and certainly with respect to climate change, experts located in institutions (such as universities) with respect to which there is less reason to suspect capture or corruption do not concur with Marohasy's strident judgements, or with the White House deletions and additions to the Environmental Protection Agency's report. Reasonable disagreement in scientific judgement relates not just to methodologies, experimental design, computer modelling, etc., but also to the location of the scientist: if there is a suspicion that ideological forces, or economic or business interests, external to scientific processes are influencing them, even though scientists themselves may sincerely deny this, then it may be reasonable to doubt the veracity of the research findings delivered by that process. This is why the research on racial categories conducted by Nazi scientists has no value. It is also why many scientists treat 'creation science' as an oxymoron.[77]

> (ii) *The Permissibility of Expertise* and *The Autonomy of Experts* will damage the potential of democratic decision-making procedures to deliver optimal solutions to problems.

This worry relates to putative efficiency costs imposed on decision making by *The Permissibility of Expertise* and *The Autonomy of Experts* in virtue of how the complications they introduce into such processes can make the problem worse by creating incentives to settle upon sub-optimal technical solutions to the problem. If *The Permissibility of Expertise* and *The Autonomy of Experts* do not transform democratic government into administrative rationalism, then the expert communities whose judgement they require policy makers to incorporate into public reasoning must not sit at the top of a decision-making hierarchy. One element in avoiding the administrative rationalism apex is to organise funding for the science base to produce a plurality of expert communities. In these conditions, such communities will inevitably compete to develop expertise in their areas. Such competition in the science base may be healthy for the state of human knowledge in the long term, but it could stimulate popular pressure to be placed on policy makers to settle on the judgements of just one set of experts for the sake of stability, especially when the problem at hand is perceived to be urgent.[78] People do not like to feel they are at risk, and they look to policy makers for decisions to reassure them; it takes a policy maker with nerves of steel to resist such pressure.

The problem is that decisions taken under these conditions can be sub-optimal from a problem-solving point of view, and can have serious opportunity costs with respect to the development of new technologies to better address the problem, given that such decisions are often to all intents and purposes irreversible. Godard gives the example of regulations on car emissions in Europe: 'these regulations practically made the use of catalytic exhaust pipes obligatory rather than allowing time for innovations in engine combustion'.[79] Another example, from Dryzek (which was UK Government policy in the 1980s), is of building tall smokestacks 'to alleviate air pollution in the vicinity of a factory, thus leading emissions to fall somewhere else and, in the case of coal-burning power plants, for sulfur dioxide to stay in the atmosphere long enough to constitute acid rain'.[80]

Again, this worry is real. With respect to climate change we see it in relation to carbon-sequestration techniques (such as those being developed by the US Department of Energy-sponsored company, FutureGen)[81] whereby CO_2 produced by power plants is captured and permanently stored by pumping it into deep saline aquifers, unminable coal seams, ageing oil wells, or depleted oil and gas reservoirs.[82] Critics claim that this technology will give a false sense of security with respect to the continuing profligate use of fossil fuels, when what is really needed is technologies (such as wind power) which have zero greenhouse gas emissions (as well as a global change in the culture of energy consumption).

> (iii) *The Permissibility of Expertise* and *The Autonomy of Experts* will affect the way in which communities of expertise operate so as to undermine public confidence in these communities.

The final worry takes it as read that experts have a key role to play in policy making and draws attention to the ways in which public perceptions of communities of experts can affect their reaction to *The Permissibility of Expertise*-type invitations to advise policy makers in a way that undermines their authority and leaves them incapable of taking up this invitation in an effective way. People tend to have faith that enough scientific research will always transform states of ignorance into states of uncertainty, and (eventually) states of uncertainty into states of knowledge of risks.[83] But this is false: sometimes more research reveals what we thought to be states of knowledge as states of uncertainty or ignorance, and (more controversially) sometimes reveals the matter to be indeterminate (i.e. permanently and intrinsically unknowable).[84] This deeply entrenched belief can generate a deep mistrust of scientists, and scepticism towards science as a whole, when the science is related to risks about which the voting public are concerned.[85] If the claims of scientists to have knowledge are used by policy makers in framing legislation, and then later – through further research – are transformed into states of uncertainty or ignorance, policy makers will lambaste scientists, and scientists know it. This likely consequence serves to make scientists – who already tend to be naturally conservative with respect to claiming states of knowledge – even more reticent with respect to delivering knowledge-claims to policy makers, and this reticence creates public scepticism about any states of knowledge that scientists might claim in the future. There is a real tension between the processes and limits of scientific research, and the expectations of the public about what such research can deliver, that threatens to emasculate scientists as experts available for consultation as specified in *The Permissibility of Expertise*.[86]

The way to address this worry – and, indeed, in some part the worries in (i) and (ii) – is through education of the public about the nature, processes, and limits of scientific research so as to enable informed interpretation of scientific judgement and reasonable expectations of scientists. Higher levels of scientific literacy may well correlate with lower (and, hence, more realistic) expectations of what scientists can do.[87]

Despite the legitimate worries just addressed, I propose that Rawls' ideal of public reason can be used to govern justificatory debate about legislative matters related to climate change once *The Permissibility of Expertise* and *The Autonomy of Experts* are appended to it. *The Permissibility of Expertise* and *The Autonomy of Experts* reflect acknowledgement that there are some disagreements related to empirical and scientific evidence the reasonableness of which non-experts cannot judge: the burdens of judgement which limit justificatory appeal in matters of fundamental justice to common sense, plain truths, and uncontroversial deliverances of science have limited mileage with respect to more complex, specialised, and controversial issues. Legislators working within the restraints of public reason must be required to seek expert advice on such matters. Public reasoning about matters of fundamental justice and constitutional essentials impose more stringent restrictions on justificatory reasons than public reasoning about matters of policy. With respect to the latter, expert considerations (within the limits of reasonable agreement as determined by the experts themselves) can serve as legitimate reasons even when reasonable non-experts reject them (witness the botanist David Bellamy's claim that climate change is 'a load of poppycock').[88]

Intergenerational justice

What should policy makers seeking justice take into consideration when addressing issues of climate change? Of key concern throughout this book will be a commitment to delivering justice to future generations, as a requirement of fairness and impartiality. If we believe – as we should – that a person's colour, religion, class, IQ, status, etc. should make no difference to what they are owed by others as a matter of justice, then the same must be true of a person's generational membership. From the point of view of justice, the point in time at which a person is born is just as morally arbitrary as the other characteristics just listed.

Rawls' principle of intergenerational justice is the just savings principle. Unlike his other principles of justice, Rawls provides no statement of it qua principle; instead, he states what the principle aims at as follows:

> The principle of just savings holds between generations, while the difference principle holds within generations. Real saving is required only for reasons of justice: that is, to make possible the conditions needed to establish and to preserve a just basic structure over time. Once these conditions are reached and just institutions established, net real saving may fall to zero. If society wants to save for reasons other than justice, it may of course do so; but that is another matter.[89]

The just savings principle is incorporated into the full statement of the principles of justice as fairness as a constraint on the difference principle.[90] This means that in justice as fairness, the achievement of intergenerational justice takes priority over the achievement of intragenerational distributive justice (more on this below). This makes sense, given the aim of the just savings principle, that is, to 'make possible the conditions needed to establish and to preserve a just basic structure over time'.[91] If the just savings principle secures (at least) a necessary condition for distributive justice, à la the difference principle, then satisfaction of this principle ought to take priority over (at least) the search for distributive justice.[92] The establishment or preservation of a just basic structure involves the achievement of a wide variety of material, cultural, technological, and collective cognitive conditions. Thus, the 'currency' of intergenerational justice ranges across 'just institutions', 'the gains of culture and civilisation', 'net investment in machinery and other means of production', and 'investment in learning and education':[93] these are the things any generation is required to save and/or preserve in the name of justice for future generations. The heart of my arguments here (crudely put) will be that because climate change threatens justice for future generations, the current generation is required to take steps now in order to avert, or minimise, this consequence.

Justice requires that each generation is required to save in order to make justice possible for subsequent generations. There are two states of generations in which this requirement applies, each of which marks the end of a spectrum along which most generations will be located. The first state is that in which full justice has been achieved: for any lucky generation in this state, the just savings principle requires that they preserve the just institutions they have achieved, or have been bequeathed, and pass them on intact to the next generation, to whom an identical requirement applies.[94] The second state is that in which justice has not been achieved, and its material base is absent: for a generation poor in this sense, the just savings principle requires that they save, but at a rate sensitive to this fact about them.[95] The just savings principle does not require that poor generations impoverish themselves even further for the sake of those to come. No generation has achieved the first state, and many have lain far closer on the continuum towards the second than the first. Thus, we can assume that in the world as we know it, all societies are required to save for the sake of intergenerational justice.

What justifies the just savings principle? An obvious first thought is that the just savings principle is the intergenerational version of the Rawls' principle of distributive justice for intragenerational contexts, the difference principle, which requires that social and economic inequalities be arranged so that they are 'to the greatest benefit of the least advantaged members of society'.[96] In *A Theory of Justice*, Rawls' starting point for thinking about intergenerational justice is indeed the difference principle: '[t]he appropriate expectation in applying the difference principle is that of the long-term prospects of the least favored extending over future generations',[97] i.e. we are to choose distributions of primary goods that make the worst-off better off than they would be under any alternative distribution, whatever generation they belong to. Following this line of thought, the intragenerational difference principle and the intergenerational just savings principle perform the same function of securing a distribution of primary goods of maximum advantage to the worst-off group within any time slice, and across all time slices, respectively.

Later on in this section of *A Theory of Justice*, however, Rawls abandons this interpretation of the just savings principle on the grounds that '[t]here is no way for later generations to help the situation of the least fortunate earlier generation'.[98] Time's arrow makes it impossible for us to implement the difference principle in the intergenerational context because the least-advantaged generation is likely to precede us and be beyond our causal reach. If 'ought' implies 'can', then the difference principle does not govern distributive justice in the intergenerational context. However, this consideration against the just savings principle qua principle of distributive justice is weak, for who is to say that the worst-off will in fact be among the generations that precede us? For all we know some future generation may be the worst-off in history, and their fate is very much in our hands. More generally, as Gaspart and Gosseries note,

> If we really care about the least well off to a significant extent, there is no good reason why, if the situation of this very least well off individual

> cannot be improved, we should remain indifferent to how the next least well off fares. If maximin advocates are ready to call for significant redistribution to the benefit of the very least advantaged, why should they not call for nearly as much redistribution to the benefit of the next least well off once the situation of the very least well off individual cannot be improved?[99]

The leximin strategy here suggested by Gaspart and Gosseries for intergenerational distributive justice is accepted by Rawls as an interpretation of what is required by the difference principle with respect to the distribution of income and wealth in the intragenerational context.[100] Simply put, in this context achieving distributive justice means making the worst-off group as well off as possible, and then the next worst-off group, and then the next, etc. If this is acceptable in the intragenerational context then, we might argue, it should also be acceptable in the intergenerational context: if the worst-off generation(s) have preceded us we are nevertheless required to focus on whatever next worst-off generations(s) might follow us. The difference principle in the intergenerational context directs us to improve the position of the worst-off who *can* be made better off, even when this group is not identical with *the* worst-off group in history, and this requires that we save for the benefit of future disadvantaged generations.

There are good reasons to avoid adopting this interpretation. We have no crystal ball and cannot be sure that, in fact, the current generation is not the next least well-off after any disadvantaged generations that have preceded us. It may be that all generations subsequent to us will be more advantaged than we are – either in virtue of living in more just societies, or in virtue of living in just societies in which the least-advantaged are more advantaged than in any previous just society – and thus to save in order to improve the position of these generations could actually stymie intergenerational distributive justice. Instead, an intergenerational difference principle may require that any current generation direct the resources they would have saved to the benefit of their own generation – either to achieve a just basic structure, or to improve the level of advantage of their own worst-off within an already just distribution of advantages, if they have achieved this.

With respect to any generation that has achieved justice, this conclusion is implied by an argument made by Gaspart and Gosseries. They refer to any such generation as in a 'steady state' and agree that, in such a state, a generation has a duty to not dis-save, i.e. not to deplete these material bases in a way that would destabilise just institutions for future generations, and not to dismantle or otherwise impair the functioning of these institutions. However, as Gaspart and Gosseries note, and as Rawls makes explicit,[101] removing the requirement to save is consistent with granting permission to save. But this permission conflicts with the rationale that motivates a move from the requirement to save to the requirement not to dis-save for generations in a steady state, as can be seen by considering the following scenario.

> Imagine that a given generation anticipates that, at the end of its existence, a surplus is likely to be transferred to the next generation, on top of the equivalent of what the current generation inherited from the previous one. If the constitution of such a surplus is likely, it should benefit the least well off members of the current generation rather than the next generation as a whole. For if each generation were to stick to this closed principle, and if we can assume compliance with intragenerational maximin by each generation, this is the scenario in which the least well off person, whatever generation she is in, will end up being at least as well off as the least well off person under any alternative intergenerational scenario.[102]

According to Gaspart and Gosseries, not only should savings in a steady state no longer be required by a commitment to intergenerational distributive justice, they should furthermore be prohibited: the ('closed') just savings principle for a steady state is 'no dis-savings, no savings'. *Pace* Rawls, any generation in a steady state is not permitted to save for 'grand projects',[103] for this would violate intergenerational distributive justice as governed by the difference principle, in virtue of how the surplus that is saved could be redistributed so as to make that generation's own least-advantaged as well off as possible.[104]

So, if the just savings principle is a principle of distributive justice, then it does not require, and indeed prohibits, savings by any generation that has achieved justice. But similar reasoning also applies to the requirements generated by the just savings principle qua principle of intergenerational distributive justice for any generation that has *not* achieved justice. Because of the uncertainty about our level of advantage vis-à-vis that of any future generation, the most we are required to do in the name of distributive justice – even when we have not achieved justice – is not to decrease the level of advantage of any future generation below that of our own, and this does not obviously require that we save for their benefit: it might require that we spend resources promoting our own intragenerational pursuit of justice while at the same time not dis-saving by ensuring that we do not put future generations at risk of harm; or, if we do, or have done so unknowingly, that we save to provide them with adequate compensation. However, there are some fundamental problems associated with making arguments that not dis-saving to avoid the imposition of risks on future generations, and saving to provide compensation to them, are requirements of distributive justice.

First, the claim that any current generation can do anything to harm any future generation – even if the latter are severely impoverished and suffering in virtue of conditions bequeathed to them by the former – is plagued by the non-identity problem. I consider this problem and solutions to it in the next section. Second, risks are not usually considered to be disadvantages addressed by principles of distributive justice,[105] and so a case for risk-avoidance in the name of these ideals is not obvious. And third, compensation is rarely justified as a requirement of distributive justice, and responds to different concerns that arise in contexts which overlay those creating concerns of distributive justice. I will argue that taking steps to avoid placing future generations at risk of harm, and making provision for adequate compensation to address risks we have already created, are in fact requirements generated by the just savings principle but *not* qua principle of distributive justice. For any current generation – whether lucky enough to have achieved justice or not – the requirements of intergenerational distributive justice are limited and may, in fact, require no savings whatsoever. However, there is a rich seam of considerations in corrective justice, and related to climate change, which will show – certainly for generations that have not achieved justice (i.e. the current generation, and all those in the foreseeable future) – that savings by people in these contexts are required. In other words, and consistent with Rawls' own account of the matter, the positive rate of savings required by the just savings principle of any generation that has not achieved justice is delivered by a commitment to intergenerational justice conceived in ways other than as distributive. I present these arguments in Chapter 4.

If the just savings principle is not justified as a principle of distributive justice, then on what grounds is it justified? By examining Rawls' solution to a sticky problem created by an appeal to impartiality in justification of the just savings principle, the details of the full justification for it can be made apparent. Rawls reasons as follows. Parties behind the veil of ignorance are to be thought of as contemporaries: they belong to the same generation (and know this) but they do not know which generation this is. Rawls calls this the 'present time of entry' interpretation of the original position.[106] Generational membership is obscured by the veil of ignorance for the same reasons that knowledge of race, religion, class, etc. are obscured, that is, in order to reflect the moral arbitrariness of these features, from the point of view of justice.

However, the present time of entry interpretation adds a dimension to parties' reasoning in the original position that is not added by their lack of knowledge of their religion, race, class, etc. Assuming, as Rawls does, that the parties in the original position are motivated by mutual disinterest (i.e. they take no interest in one another's interests), for them to agree on a principle of just savings would be for them to gamble with their interests in an impermissible way: it would be for them to gamble on being in a generation that benefits from the savings of previous generations, rather than in a generation that saves in a way that will only benefit future generations. The mutual disinterest of parties in the original position, in combination with their conception of themselves as 'trustees … responsible for citizens' (unknown) determinate and complete good',[107] means that they focus on improving the position not of those who would benefit most from the principles they choose, but rather on those who would be least advantaged by these principles, because the principles they choose as trustees must be justifiable to all those for whom they act in trust.[108] In the case of intergenerational savings, those who would be least advantaged by a just savings principle are 'the least fortunate first generation'.[109] Because time's arrow points in one direction – because '[w]e can do something for posterity but it can do nothing for us'[110] – if the least-fortunate first generation are required to save, then they make sacrifices for which they receive no return: only future generations will benefit from their savings. Because the position of the least-fortunate first generation would be improved if they were not required to save, it follows that – given the present time of entry interpretation, and holding fixed the motivation assumption of mutual disinterest – the parties in the original position reject the just savings principle.[111] The worst-off generation will be made as well off as possible if it does not save for future generations. Call this the 'Hang Posterity' problem.

In *A Theory of Justice* Rawls approaches the Hang Posterity problem as follows. Preserving the present time of entry interpretation of the original position, parties (1) (know that) they 'represent family lines … who care at least about their more immediate descendants'[112] (i.e. Rawls changes the motivation assumption for the original position), and (2) are required to adopt a principle of savings 'such that they wish all earlier generations to have followed it'.[113] These modifications to the original position mean that the parties consider not only relations of justice between persons at any given point in time, but also such relations as they stretch backwards and forwards in time. Rawls describes their reasoning as follows:

> [The parties] try to piece together a just savings schedule by balancing how much they would be willing to save for their more immediate descendants against what they would feel entitled to claim of their more immediate predecessors. Thus imagining themselves to be fathers, say, they are to ascertain how much they should set aside for their sons and grandsons by noting what they would believe themselves entitled to claim of their fathers and grandfathers.[114]

Thus, according to Rawls, the parties do not choose a just savings principle in the name of intergenerational distributive justice. Rather, the just savings principle is the institutional expression of the natural concern parents have for their children, and their children's children, and the natural expectations offspring have of their parents. The reason for any generation to save for future generations no longer expresses impartial equality between generations, but rather natural bonds of kinship, which are divested of any partiality through the present time of entry interpretation, and translated into expectations about the resources one generation ought to bequeath to the next.

Prima facie this provides an attractive vision of why we should commit to intergenerational justice: we love our kids and don't want them to suffer. But there are at least two good reasons to be wary of the 'family lines' approach.

(1) Unless we are to build parenthood into the conception of the person upon which justice is premised – perhaps by adding this status to the list of things people know about themselves when choosing principles in the original position – then the family-ties approach is stymied. The point of this list is to specify the features of persons in virtue of which they are all equally deserving of justice. But why think that being a parent is such a feature? Unlike all the other

features of parties in the original position of which they are aware – possession of the two moral powers, knowledge that they have a conception of the good without knowing its content, knowledge of the list of social primary goods of which they prefer more to less – this feature is not universally shared by all persons insofar as they claim justice, and the lack of it does not render a person unable to participate in social cooperation on terms of reciprocity.

(2) The 'family lines' amendment to the original position is unsatisfactory both on Rawlsian and on its own terms. It is out of kilter with the spirit of the veil of ignorance insofar as it takes a highly culturally specific expression of familial bonds, i.e. inheritance, to be the definitive institutional reflection of the sentimental ties that constitute these bonds. The impartiality of the veil of ignorance is supposed to deliver principles that enable justice-based critique of such contingent and historically particular practices; building such a practice into the architecture of justice makes such critique impossible.[115] Furthermore, making kinship the basis of intergenerational duties undermines the status of justice as the first virtue of social institutions,[116] whereby justice is supposed to regulate kinship relations, rather than be derived from, limited by, or reflect the content of them.

If we want to avoid this result, there are at least two ways forward. First, *pace* Rawls, we could deny that the Hang Posterity problem is insurmountable without adjustments in assumptions about the motivation of parties in the original position; second, we could replace the 'family lines' motivation assumption with a different one that does not threaten to undermine the impartiality of the original position. The first approach is developed by Gaspart and Gosseries; the second – I shall argue – builds upon the Gaspart-Gosseries strategy by making good a defect at its heart through adjusting the motivation assumption in a way that does less violence to the original position, and which is made possible by Rawls' comments about the justification of the just savings principle in *Political Liberalism* and *Justice as Fairness*. It is in this second adjustment that the most promising solution to the Hang Posterity problem is to be found, and on which an account of the motivation problem with respect to climate change will be built in Chapter 6. However, let me lay out the Gaspart-Gosseries argument as a prelude to this.

Gaspart and Gosseries argue that the reason parties in the original position have for endorsing the just savings principle is exactly the same as the reason they have for rejecting a basic structure which denies them basic liberties in order to deliver greater income and wealth to the worst-off. The priority the parties in the original position give to liberty over other goods supplies the reasons in each case. Here is the key paragraph:

> [O]n the one hand, accumulation can be justified in the name of reaching a level of wealth enabling a society to minimally guarantee the protection of basic liberties. On the other hand, accumulation would in principle violate maximin once applied intergenerationally. Yet, given the lexical priority of the principle of equal liberty over the difference principle, Rawlsians are able to defend such an accumulation phase [i.e. a positive rate

> of savings – CM]. Admittedly, not the *same* people will both suffer from the violation of maximin (earlier generations in the accumulation phase) and benefit from society's ability to protect basic liberties (generations beyond the end of the accumulation phase). However, this is also the case in the intragenerational case ... those who could benefit from redistribution (the least wealthy) will not necessarily be the same as those benefiting from a ban on significant violations of freedom of expression (*ex hypothesi*, the most wealthy members of society).[117]

This interpretation makes no appeal to family lines or kinship. Does it address Rawls' worry about how time's arrow would lead parties in the original position to hang posterity? The argument is that the justification of sacrifices by the worst-off by reference to the priority of liberty in the intragenerational case also applies to such sacrifices in the intergenerational case. The success of this argument depends on an analogy between the group of people in the intragenerational context, and the group of people in the intergenerational context, who are made less well off than they would be otherwise by distributions of – let us say for simplicity – economic goods according to principles over which the principle of equal liberty takes priority. In the intergenerational case this group is the worst-off in the least-advantaged – say the first – generation (call them 'the serfs'); in the intragenerational case this group is the worst-off in any time slice (call them 'the have-nots'). However, there is the following crucial difference between the two groups. *Ex hypothesi*, the have-nots enjoy a just distribution of a fully adequate scheme of basic liberties (albeit at the cost of fewer economic goods than they would otherwise have): the sacrifices they make are required by the existence of just institutions which they have before their eyes, and the benefits of which they enjoy. However, the serfs do not have these institutions before their eyes, and do not enjoy the benefits of them: the sacrifices they make are in order that future generations can achieve these benefits. And knowing this justifies the parties' rejection of the just savings principle and their decision to hang posterity.

This disanalogy means that the strong arguments for the priority of liberty that support the sacrifice of economic benefits in the intragenerational case do not obviously carry over to support the case for savings in the intergenerational case. However, if the analogy can be preserved by showing how the serfs and the have-nots both enjoy the same good in virtue of sacrifices they make of economic goods, despite their different lived experiences of justice, then perhaps the argument can still go through.

The key question is: what, if anything, do the serfs and the have-nots share, despite the fact that the latter enjoy a fully adequate scheme of liberties distributed by a set of just institutions that the former do not? Some later remarks by Rawls suggest an answer to this question.

> [W]henever there is a shared final end, an end that requires the co-operation of many to achieve, the good realized is social: it is realized

> through citizens' joint activity in mutual dependence on the appropriate actions being taken by others. Thus, establishing and successfully conducting reasonably just (though of course always imperfect) democratic institutions over a long period of time, perhaps gradually reforming them over generations, though not, to be sure, without lapses is a great social good and appreciated as such. This is shown by the fact that people refer to it as one of the significant achievements of their history.[118]

A just society is an end that requires the cooperation of many to achieve, and thus qualifies as a social good. It is this good, I propose, that the have-nots and the serfs have in common: both groups, albeit through different forms of activity, engage in cooperation through time to achieve a just society. For the serfs, their enjoyment of the social good of cooperation to achieve justice takes the form of saving (at the appropriate rate) so that later generations have resources sufficient to create just institutions. For the have-nots, their enjoyment of this social good takes the form of sustaining, and passing on, just institutions. Hence it is incumbent on the have-nots to accept, if necessary, fewer economic goods for the sake of the priority of liberty, because the institution of this priority rule is part of what makes a society just. Preserving the present time of entry interpretation, and keeping fixed the motivation assumption of mutual disinterest, it is nevertheless the case that parties would agree to a positive rate of savings for generations lacking resources sufficient for them to achieve justice for themselves in virtue of the significance they attach – whatever their generation – to the social good of cooperation to achieve justice: 'all citizens give high priority to the end of cooperating politically with one another on terms that the representative of each would endorse in a situation in which they are all fairly represented as free and equal, and reasonable and rational'.[119] Parties in the original position would *not* agree to hang posterity on the grounds that those they represent could be serfs, because to do so would be to deprive these people of a great good: the social good of cooperation to achieve a just society. As we shall see, this has far-reaching consequences for political thinking about climate change, especially with respect to understanding the sense in which people not motivated to tackle climate change for the benefit of future generations exhibit a moral failing.

For now, let me anticipate and reply to three objections to the idea of intergenerational cooperation to achieve justice as a social good.

(1) Granting that saving by the serfs, and foregoing greater economic benefits for the sake of equal liberty by the have-nots, delivers to each of them the same good of cooperation to achieve a just society, there remains a disanalogy between the groups: the have-nots actually live in a just society whereas the serfs do not.

In reply, as I stated in my objection to the Gaspart/Gosseries argument, it is true that this disanalogy holds. Because the have-nots are lucky enough to live in a time of justice (or of greater justice than the serfs) they enjoy (or at least have available to them) what Rawls elsewhere identifies as two further aspects of the good of a just society: phenomenological goods (the experience of living in a just society) and objective goods (the delivery of justice to members).[120] But the claim I am defending is not that the serfs and the have-nots are identically well off; clearly they are not. The have-nots are lucky to be living later than the serfs. Rather, the claim being defended is that parties in the original position have justice-related reasons for imposing requirements on *both* the serfs and the have-nots that they accept, if necessary, fewer economic goods (either through saving, or through prioritising liberty) for the sake of achieving a just society. The form of activity for each group that might be so required will differ: the serfs are required to save, whereas the have-nots are (*inter alia*) required to institute and preserve the priority of liberty. And members of the have-nots, even though economically worst off in their society, will receive (phenomenological and objective) goods related to living in a just society that are lacked by the serfs, who forego economic benefits but do not experience justice. Nevertheless, there is commonality in at least one benefit delivered to both groups who make such sacrifices for these reasons: receipt of the social good of cooperation to create a just society.

(2) The serfs can have no certainty that their saving will achieve justice: their cooperation may turn out to be futile if subsequent generations squander their inheritance or abandon transhistorical cooperation to achieve justice.

In response, this is true, but it does not defeat the argument, which is supposed to provide parties in the original position with justificatory reasons for imposing requirements to save on the serfs, rather than provide the serfs with motivating reasons for doing what they are required to do. In other words, the argument so far is located in ideal theory, where we assume that people comply with the requirements of justice.[121] Of course, in the non-ideal real world of partial compliance the serfs may well be suspicious of the good intentions of their successors, and may decide on that basis to hang posterity. But if the argument I have made so far holds good, any such decision by a generation so situated would be unjust. Parties in the original position will know that no generation has a crystal ball, nor can bind the hands of its successors. But this does not detract from the significance they attach in the original position to the social good of cooperation to achieve a just society: even if such cooperation by one generation failed to achieve its objective in virtue of the actions of later generations, that cooperation is not thereby rendered worthless (as we tell children, 'it's the taking part that counts'). To borrow some terminology from Janna Thompson's illuminating discussion of intergenerational justice, members of each generation have 'lifetime-transcending interests' that generations subsequent to them will continue the project of cooperation to achieve justice.[122] But it is in the nature of the case that the most that any generation can do to secure these interests is to contribute to the project themselves.

(3) The serfs may have no awareness that they are receiving the social good of cooperation to achieve a just society, and it stretches credulity to claim that a good received of which one has no awareness provides justifying (let alone motivating) reasons to make sacrifices to the extent that could be required of the serfs.

In response, it does not stretch credulity to make this claim: many goods are received irrespective of the knowledge of that fact by the receiver. It is a good to be protected, to be lucky, or not to be lied to: but a person can be in receipt of these goods without any knowledge of the fact. The serfs, and indeed the have-nots, may have no sense of themselves as cooperating to achieve a just society; nevertheless, they do, and they receive the good associated with this activity.

In sum, the argument that addresses the Hang Posterity problem by providing parties in the original position with a reason for imposing a requirement to save on the serfs is that such saving delivers to them the important social good of cooperation to create a just society.

The account of intergenerational relations as cooperative delivers justifying reasons to support the just savings principle. However, reasons which justify but are not fit to motivate make for an impoverished political justification. Successful justifying reasons must be fit to motivate people to act for these reasons; they must be action guiding. Justifying reasons must be motivationally adequate because political justification is a practical activity: it ultimately aims to convince people that either they ought to restrain themselves according to its principles, or that the coercion used to restrain those who do not restrain themselves is legitimate. A reason which was not fit to motivate people to act could not be a practical reason. The construction of political justifications must have an eye both on the question of whether people *could* act for the reasons it offers, and on the question of what it would take for these reasons to become reasons upon which people *do* act. In Chapter 6 I shall give an account of how the reasons that justify doing intergenerational justice can also generate motivating reasons; in the intervening chapters I shall reflect on what the intergenerational scope of justice means for thinking about climate change within a liberal framework. For now, though, one more issue must be addressed by way of preliminaries.

Interlude: the non-identity problem

It is now time to address what has become the bête noire of any theoretical approach that includes future generations as recipients of justice: the non-identity problem. I shall present the problem in general terms,[123] and then show why the Rawlsian approach escapes it. My intention is to discuss the problem in order to set it aside. Although it presents a real philosophical puzzle about how some approaches to justice can conceptualise harm-avoidance principles with intergenerational scope, its dominant place in debate about intergenerational justice does no favours to philosophical contributions to such debates, and – in my opinion – has become a serious hindrance to progress in the face of real and pressing political problems.

In a nutshell, the non-identity problem is this: because any future people who come into existence do so only as a result of the decisions we take in the present, they can have no complaint against us that we have harmed them, because if we had acted differently they would not have existed at all, and it is better to exist than not to exist. The non-identity problem arises from the fact that each person's existence depends upon a unique moment of conception determined by countless prior events and facts: the location and state of mind of the mother and father, their health and levels of nutrition, the circumstances of their meeting and their identities, etc. Each of these events and features of the mother and father can be affected, and are framed, by features of the world in which the mother and father live. For example, the political regime in which they live affects their opportunities, education, work, and leisure in a way that can affect the chances of their meeting; the way in which they were raised can affect the point at which intercourse between them takes place; and even natural events such as thunderstorms and sunny days can make a difference to the sperm and egg that combine at the moment of conception. And all of this is true of the mother's and father's parents, and their parents, etc. What this means is that the choices each of us make now – from the most mundane choices of everyday people, to the most momentous political decisions of statesmen – make a difference to the identity of the people who will exist in the future. A different set of choices today will cause a different person – or no person at all – to come into existence next year, or the year after that, or the year after that, etc.

The non-identity problem is that if the existence of future people is dependent on our present choices then future people can have no complaint against us that we have harmed them through these choices, because if we had acted differently they would not have existed at all, and it is always better to exist than not to exist: there is no way other than the way we in fact acted which would have caused the existence of the future people with the putative complaint against us. An obvious rejoinder to the non-identity problem is that it is not always better to exist than not to exist: there are lots of fates worse than death. However, this elides two senses of the claim 'it is always better to exist than not to exist': (a) that it is always better to exist than never to have existed, to have come into being than never to have been at all; (b) that it is always better to exist than to stop existing, to cease to be. Claim (b) is clearly false: plenty of suicides make sense, and there is a strong moral case for voluntary euthanasia. However, the non-identity problem operates with (a) not (b): even for people who would be better off ceasing to exist, it is always better that they exist than that they never come into being at all.

The non-identity problem putatively stymies the articulation of principles of intergenerational justice, insofar as these principles aim (at least in part) at avoiding the infliction of harm on future people, because if it is not possible for us to inflict these harms, then it is not possible for us to act unjustly towards future generations. I shall argue shortly that the non-identity problem does not stymie a Rawlsian approach. To get to that point, consider what is being claimed in (a), the truth of which is required for the non-identity problem to stand. Claim (a) asks us to compare two states: one in which a person exists, and one in which she does not. But how are we to conceive of a person in a state of non-being in order to make this comparison? A person in a state of non-being is not a person about to be. A non-existent person is nothing at all. For that reason the comparison upon which the truth of (a) depends cannot be made.[124] Putting things another way, we can see that it is not correct to think about what we owe future generations as follows: future generation F, i.e. the set of people P1, P2 … P6bn, is owed justice by us. This formulation is vulnerable to the non-identity problem because it treats the term 'future generation F' as a rigid designator, that is, as a term that designates the same object in all possible worlds. The mistaken line of thinking is this. In some worlds F does not exist (its members are merely possible). But in this world, F exists, and its existence is a result of what we do. And if existence is better than non-existence, then we have benefitted people P1, P2 … P6bn who comprise F. But it is a mistake to frame our thinking about future generations as if there are multiple – perhaps an infinite number of – sets of possible people waiting to come into existence, and our choices now will set a path to the future in which one of these lucky sets of people comes to be, and the other unlucky ones remain merely possible. Whoever will be in the future, will be, and their coming into existence as a result of choices we make is not a benefit we bestow upon them, but is rather a necessary condition for benefits (or harms) that could be caused to them by our actions now. Putting things another way, a person existing in 2200 cannot meaningfully say, 'Thank goodness I exist! I would have felt so much worse had I not': her existence is a condition of her being anything at all, and if this condition does not obtain there is no 'she' to be worse off.[125]

The correct way to think about what we owe to future generations is this: future generation F, whoever is in it, is owed justice by us. In other words, the term 'future generation F' should be treated as a non-rigid designator. F could be comprised of P1, P2 … P6bn, or it could be comprised of an entirely different set of people. It is true that what we do now will affect the identity of members of F, that is, will affect who it is that actually comes to be. But what matters is that, whoever comes to be, what we do now delivers justice to them. And the reason for this is, as explained in the last section, that the point in time at which a person – any person – is born is as morally irrelevant to her status as a recipient of justice as is her sex, colour, religion, etc. Just as we can say that a person, whatever her skin colour, deserves justice, so we can say that a person, whatever her generation, deserves justice.

Jeffrey Reiman makes this point by arguing that what matters from the point of view of intergenerational justice conceived in Rawlsian terms is that the 'properties [future people] are born with or into' make a difference to whether their morally relevant interests are respected, and it is these properties to which principles of intergenerational justice speak, regardless of the particular identities of the persons who will have these properties.[126] The morally relevant interests of future people – whoever they are – are the same as present people's (although what future people require in order to pursue these interests may be radically different from the requirements of present people), and it is this that is modelled by the original position. As he puts it,

> [F]uture people are represented in the original position as the people who will exist in the future whatever particular individual traits they turn out to have, *indeed, whatever particular individuals they turn out to be.* From this it follows that what we owe to future people is not owed to them simply in virtue of the particular individuals they will be. We owe it to them simply in virtue of the fact that they will be people … .[127]

This is the right way to think about intergenerational justice, and makes the perspective of what we owe to future people, rather than what they (as the particular people they will be) can claim from us, the appropriate one to adopt. That thinking about intergenerational justice from the original position escapes the non-identity problem is yet another reason to prefer the Rawlsian stance over others. Because this is the stance adopted throughout the rest of the book, I shall say nothing more about the non-identity problem from here on in.

Ideal theory, background justice, and the circumstances of justice

Before addressing, in subsequent chapters, some questions of justice raised by climate change it is worth noting the nature of the theorising about justice that will be undertaken here. As Rawls is the major influence for this task, let us start with his distinction between ideal and nonideal theory. Rawls primarily draws this distinction in terms of strict and partial compliance with the requirements of just institutions. The aim of ideal theory is to '[e]xamine the principles of justice that would regulate a well-ordered society … [wherein] everyone is presumed to act justly and to do his part in upholding just institutions'.[128] In contrast, nonideal theory

> 'studies the principles that govern how we are to deal with injustice. It comprises such topics as the theory of punishment, the doctrine of just war, the justification of the various ways of opposing unjust regimes … [a]lso included here are questions of compensatory justice and of weighing one form of institutional injustice against another'.[129]

As Rawls acknowledges, 'the problems of partial compliance theory are the pressing and urgent matters'.[130] For this reason, some contemporary political theorists have little patience with, and indeed reject, ideal theory à la Rawls (or anyone else) as a useful starting point in political thinking.[131] However, it is a great mistake to do so. When done well, ideal theory provides a blueprint for thinking about the values and principles of a perfectly just society. Even if such a blueprint makes no difference whatsoever to the behaviour of the perpetrators of injustice in the actual world, it still matters enormously that we understand why what these perpetrators do is unjust. Without at least a partial conception of the ideally just society such understanding will not be possible. Moreover, not only will understanding not be possible, but also action-guiding, practical reasons to behave in this way rather than that way so as to move closer to justice will not be possible. As Adam Swift puts it, 'only by reference to philosophy – abstract, pure, context-free philosophy – can we have an adequate basis for thinking about how to promote justice in our current, radically nonideal, circumstances'.[132]

Notwithstanding the erroneous rejection of ideal theorising per se, Rawls' account of it can be improved upon. Ingrid Robeyns offers a useful account of the aims and role of ideal and nonideal theory that goes well beyond the simple distinction between full and partial compliance. She argues that nonideal theory ought to provide an 'effective bridge' between ideal theory, and action design and implementation.[133] Whereas ideal theory provides a (often no more than partial) blueprint of the perfectly just society, nonideal theory '[offers] us the theoretical foundations for figuring out what we have to do in order to move closer to that society' (through guiding our comparisons of the justice of different social states – using ideal theory as a touchstone – and our consideration of the actions required to realise more justice),[134] in order that social scientists can give an account of the feasibility constraints on the social states under consideration, of the possible unintended consequences of policies which aim at the creation of these states, and of various implementation strategies for policies under consideration.[135] In sum,

> Ideal theory surely has an important role to play, but is far from sufficient, since there is no straight bridge between ideal theory and guidance of actions. We need nonideal theory, and we have to examine whether we can use the resources offered by ideal theory to work out a set of principles for how to proceed in situations in which idealizing assumptions are not met.[136]

It is to this conception of nonideal theory that I take the arguments of this book to belong. The arguments I make for the precautionary principle (Chapter 3), compensation for future generations for the harms of climate change in the name of corrective justice (Chapter 4), principles of triage for climate change catastrophes (Chapter 5), and a strategy for motivating members of the current generation to do climate change justice (Chapter 6) are all made in terms of the values of justice as expressed in (Rawlsian) ideal theory, but relate to political problems which have arisen (or will arise) in virtue of persons' failure to do justice, and are intended to provide normative platforms for discussion by social scientists of policies and implementation strategies fit to realise them. In this way the hope is that the arguments of this book could contribute towards moving us closer to a state of justice vis-à-vis the intergenerational political problems created by climate change (although, as stated in Chapter 1, there is no expectation that this will happen).

Although, as I have said, the arguments of this book are not doctrinaire, there is one aspect of Rawls' account of ideal theory which will emerge and re-emerge as a theme throughout all the discussions. It can be found in the following passage:

> The two principles [of justice] ... specify an ideal form for the basic structure in the light of which ongoing institutional and procedural processes are constrained and adjusted ... We need such an ideal to guide the adjustments necessary to preserve background justice ... [E]ven if everyone acts fairly as defined by the rules ... the upshot of many separate transactions will eventually undermine background justice. This is obvious once we view society, as we must, as involving cooperation over generations.[137]

For Rawls, ideal theory specifies the principles to which a just basic structure must conform if it is to preserve background justice, without which just social cooperation is not possible, even when all involved have the best of intentions and are committed to realising justice.[138] Without the institutions of a just background structure operating as a backdrop to, and a constraint on, the myriad choices of individuals as they interact in the various arenas of human life, the outcomes of these choices would eventually accumulate to create a background of injustice, whereby any initial equal freedom would be eroded. Rawls' example is of how voluntary market transactions can undermine a fair distribution of wealth.[139] A key touchstone to which I shall return time and again in what follows is the idea of what preserves background justice: for Rawls, this is a just basic structure. But sometimes external circumstances can make it impossible to erect and/or maintain the institutions of a just basic structure, especially when the sacrifices required of people in order to achieve a just basic structure are too great. A key contention throughout the book will be that climate change threatens to destroy the external circumstances necessary for the institutions of the basic structure to preserve background justice, and that this generates a weighty reason (drawn from ideal theory) to address climate change in nonideal theory, not just for the benefit of future generations, but also for the benefit of all members of the current generation qua participants in the project of pursuing justice over time.

3

PRECAUTIONS

> [T]he existing estimates of [the] social cost [of climate change] are based on IPCC studies that so far have not included many … irreversible positive feedbacks … So nobody has yet even asked what price should be attached to a century-long drought in the American West, or an enfeebled Asian monsoon, or a permanent El Niño in the Pacific, or a methane belch from the ocean depths, or a collapse of the West Antarctic ice sheet, or sea levels rising by half a yard in a decade. Though, on reflection, these are perhaps questions not best answered by accountants.[1]

Introduction: catastrophes and tipping-points

In this chapter I shall make a case for adopting the 'precautionary principle' in policy making addressed to climate change by reflecting on the worst-case scenarios associated with a failure to adopt this principle, and on the obligation of the current generation to attempt to minimise the risks of these worst-case scenarios coming to pass for future generations, in the name of intergenerational justice. I shall also make some more specific suggestions as to how, exactly, the precautionary principle could be operationalised by policy makers, which connect up with the general account given in the last chapter of the role of experts in debate about climate change policy governed by public reason.

The basic science of climate change – our understanding of the way in which greenhouse gases warm the planet – is well established and understood. However, experts still have limited understanding of the nature and significance of various positive and negative feedback effects that could respectively accelerate or slow climate change, and there is much disagreement among them about the accuracy and reliability of the models positing these effects, and/or the existence of data sets adequate to support their prediction.[2] In particular, the possibility of powerful positive feedbacks is increasingly being taken by many experts to undermine the 'gradualist paradigm' in thinking about climate change, most notably as adopted by the IPCC.[3] According to this paradigm the changes constitutive of climate change are cumulative, smooth, and incremental, which makes them easier to predict and plan for. However, challenges to this paradigm have it that climate change might happen through a combination of gradual changes and some very abrupt shifts, as we pass various 'tipping-points'. These events are points of no return, beyond which positive feedbacks causing runaway climate change create a world not represented in any IPCC scenarios,[4] and beyond the scope of much scientific imagination to date. The possibility of passing tipping-points on the way to climate change catastrophe has led some normally cautious scientists to adopt the language of Armageddon in their attempts to get these possibilities into the public debate about climate change; for example, NASA scientist James Hansen describes the current state of affairs as 'a planetary emergency'.[5]

There are numerous significant tipping-points, and new ones are emerging all the time. Very much simplified, some frequently mentioned tipping-points are the shutdown of the Atlantic conveyor (which distributes heat to Western Europe) as a result of melting ice caps, possibly causing another ice age; rapid and large rises in sea level (again caused by massive melting of ice sheets and glaciers) fit to swamp London, Sydney, New York, Tokyo and most of the US seaboard, as well as low-lying Bangladesh, and many (probably already doomed) small island states; the melting of the permafrost in Siberia, causing the release of huge stores of methane (a greenhouse gas twenty times more potent than CO_2) from the peatlands below (present estimates are that the melting permafrost is releasing 100,000 tons of methane a day, which has a warming effect greater than all the US's daily greenhouse gas emissions);[6] a rapid decrease in the Earth's albedo as a result of melting ice and decreased snow cover, causing a drastic reduction in the Earth's capacity to reflect sunlight back into space, and leading to runaway warming.[7] Many climate change catastrophes would cause runaway climate change by intensifying positive feedbacks that would accelerate it. Runaway climate change is the phenomenon whereby certain events heighten the speed and magnitude of climate change, causing further such events, with the result that climate change would exceed human capabilities to mitigate and/or adapt to it. If runaway climate change occurs, humankind will be in a position to do no more than observe the process, and hope for minimally bad impacts.

Let me note four things about runaway climate change subsequent to passing a tipping-point. First, such climate change has happened many times before in the Earth's history. For example, many scientists believe that 12,800 years ago as a result of deglaciation a massive freshwater lake – Lake Agassiz – emptied into the North Atlantic. This disrupted the warmer currents moving northwards and, within a matter of decades (perhaps less – see below), plunged the world into an ice age (the Younger Dryas) that lasted 1,300 years.[8] Many climate experts believe that such lurches between extreme states of glaciation and deglaciation are the norm, and that the relative stability of the climate in the present interglacial period, since the end of the Younger Dryas 8,200 years ago, is an anomaly.

Second, runaway climate change often leads to, and feeds off, catastrophe on a scale not known in recorded history. *The Oxford English Dictionary* defines catastrophe as 'a sudden great disaster'; Webster's is more expansive in defining it as 'a momentous, tragic, usually sudden event marked by effects ranging from extreme misfortune to utter overthrow or ruin'. 'Catastrophe' is clearly a loaded term, and as such is often eschewed by scientists (it does not appear in any IPCC Reports, for example), many of whom advocate instead that the processes and events of runaway climate change be referred to as 'dangerous climate change'. Notwithstanding the reticence of scientists, the consequences of passing a tipping-point for life on Earth mean that if the runaway climate change that would ensue does not count as a catastrophe, then nothing does, so I shall continue to use the term.

Third, climate change catastrophes can be sudden. To return to the Younger Dryas, evidence from Greenland ice cores shows that the warming that took the world out of that period and into the Holocene took place within a decade. However, as Fred Pearce reports, one respectable scientist in the team that made this claim is willing to go further. Richard Alley states that '[m]ost of that change looks like it happened in a single year. It could have been less, perhaps even a single season. It was a weird time indeed.'[9] So, the increasing doubts of experts about the adequacy of the gradualist paradigm for thinking about climate change should be supplemented with the thought that runaway climate change may happen fast not just in geological time, but in human time too.

Fourth, the state of knowledge with respect to tipping-points and climate change catastrophes is such that, with respect to many of them, experts do not know their proximity to us in time. With respect to many catastrophic climate change events, we could be on the verge of the tipping-points that make them inevitable. A good example is the release of vast quantities of methane from clathrates lying below the world's warming oceans.

To provide focus, it is this catastrophic climate change event to which I shall limit discussion, and which is related to a set of frightening positive feedbacks to which scientists have awakened only recently. Fifty-five million years ago the Earth experienced a period of extreme and accelerated warming called the Paleocene-Eocene thermal maximum (PETM), which triggered the largest mass-extinction event since the 'Great Dying' 251 million years ago (in which 95 per cent of species went extinct). Many scientists believe that the PETM was largely caused by the release of trillions of tons of methane from beneath the oceans. This methane had been trapped in frozen, honeycombed sediments called 'clathrates', and was released as a result of global warming (probably caused by solar activity). At present, 1 to 10 trillion tons of methane clathrates exist beneath the world's oceans, and could be destabilised by the rapidly warming oceans. The release of even a fraction of this methane could be catastrophic in at least two ways. First, it could cause massive underwater landslides that would create tsunamis comparable in size with, or larger than, the one that hit Japan in 2011. Indeed, scientists now believe that the 'Storegga slip' 8,000 years ago – which occurred off the coast of Norway and caused a tsunami of 40 feet in Norway, 20 feet in Scotland, and 65 feet in the Shetlands – was caused by the explosive release of methane from a destabilised clathrate. Second, the release of methane on this scale could cause runaway climate change from which it would be impossible to recover, given existing levels of anthropogenic warming.[10] It is not science fiction to claim that in these conditions temperature increases could exceed 6°C. In this scenario the majority of life on Earth, perhaps including *Homo sapiens*, could go extinct.[11] Call this scenario 'Methane Nightmare'.

I shall argue that a precautionary approach ought to be adopted by policy makers addressing climate change, in virtue of the possibility that climate change could cause us to pass key tipping-points beyond which positive feedbacks are activated that could cause runaway and catastrophic climate change. I shall present arguments for the precautionary principle as the approach to climate change catastrophes that should be adopted by policy makers concerned (as they should be) with intergenerational justice, and I shall understand intergenerational justice in broadly Rawlsian egalitarian terms. What I mean by this, in part, is that it is appropriate to adopt Rawls' conceit of the original position when thinking about intergenerational justice. The impartiality modelled by this method for choosing principles of justice is required not only between members of the same generation, but also across members of different generations. Rawls incorporates this commitment into his hypothetical social contract by stipulating that intergenerational justice is to be governed by his 'just savings' principle, as outlined in Chapter 2.[12] Parties behind the veil of ignorance are to be thought of as contemporaries: they belong to the same generation (and know this) but they do not know which generation this is. He calls this the 'present time of entry' interpretation of the original position.[13] In this situation, parties would choose principles of social cooperation according to which the worst-off group is made as well off as possible and, for Rawls, this holds with respect to intergenerational as well as intragenerational justice.

In this chapter I shall distinguish between two Rawlsian arguments for taking precautionary action against the worst outcomes of climate change. The first turns on reflections about the injustice of the distribution of advantage across generations that could be caused by failing to take precautions against climate change catastrophes, and the second on how a failure to take precautions puts future generations at risk of suffering unbearable strains of commitment that would make the pursuit of justice impossible for them. I shall show that although both arguments provide compelling grounds for taking precautionary steps to avert climate change, the second argument is more powerful than the first insofar as it transcends considerations of intergenerational distributive justice which – as we saw in the previous chapter, and will see again in Chapter 4 – have a limited, and at best problematical, role in arguments for addressing climate change for the sake of intergenerational justice.

The precautionary principle

One feature of climate change catastrophes that makes policy making with respect to them so difficult is that they are uncertain. In one basic sense, the meaning of 'uncertainty' is clear: that which is uncertain is not certain. Given that nothing in the future is certain, and given that all policy is future oriented, uncertainty (in this basic sense) is integral to the context of every policy decision. Uncertainty in this sense does not pose a special problem for policy makers because it is consistent with a future event being uncertain that it is assigned a probability of happening (indeed, probability is a way of expressing uncertainty), and once a policy maker has this information she can perform a standard risk assessment to guide her thinking about the event by multiplying its probability by its costs or impact.

This is not the sense in which many climate change catastrophes are uncertain. Rather, although many climate change catastrophes are known to be possible (often because they have happened in the past), the processes that cause them are so poorly understood that it is not possible to assign a precise numerical probability to them, and thus a risk assessment cannot be made.[14] Clive Spash's typology of uncertainty as 'weak' and 'strong' is helpful here.[15] Weak uncertainty exists when outcomes and their probabilities (understood either as objective features of the world, or as subjective insofar as derived from persons' preferences) are known. The context for much policy making is weak uncertainty, and the dominant approach to policy making in this context is risk assessment. Strong uncertainty exists when outcomes are unknown and/or unpredictable, either because the outcomes are indeterminate, or because the state of knowledge with respect to these outcomes precludes the assignment of a probability to them.

The strong uncertainty of many climate change catastrophes renders them an elephant in the room in climate change policy debates. Of course, some greenhouse gas reduction targets – for example, the UK's commitment to cut CO_2 by 80 per cent by 2050 (relative to a 1990 baseline, as per the Climate Change Act 2008) – can be read as addressed to climate change catastrophes just in virtue of the fact that *any* reduction in greenhouse gases can be seen as an attempt to mitigate climate change, catastrophes included. But such policies are certainly not justified by direct reference to mitigating climate change catastrophes; and a cut of 80 per cent (let alone the paltry Kyoto targets, and putting to one side the shameful failure of the international community to agree legally binding targets for CO_2 reductions at COP-15) is unlikely to do the job. Furthermore, as already noted, the bibles of policy makers addressing climate change – the IPCC Reports – do not include climate change catastrophes in their scenario analyses, and work largely within the gradualist paradigm.

The unavailability of risk assessment as a method for policy making with respect to strongly uncertain climate change catastrophes has led an increasing number of thinkers (even those generally sympathetic to cost-benefit analysis, of which risk assessment is an example)[16] to give serious consideration to the precautionary principle as a guide to policy making for such events.[17]

Before moving on to consider philosophical aspects of, and problems related to, the precautionary principle it is worth considering the limited impact it has had in the world of politics and policy, and why this might be so. The precautionary principle has been incorporated into some key political documents with an international focus, although the statement of it often differs across these documents, and sometimes it is simply mentioned in name only. For example, the Montreal Protocol on Substances that Deplete the Ozone Layer (1987) simply states that: 'Parties to this protocol … determined to protect the ozone layer by taking precautionary measures to control equitably total global emissions of substances that deplete it', and gives no elaboration of when such measures are appropriate and what they are. In contrast, the Framework Convention on Climate Change (1992) states that:

> The Parties should take precautionary measures to anticipate, prevent, or minimise the causes of climate change and mitigate its adverse effects. Where there are threats of serious or irreversible damage, *lack of full scientific certainty* should not be used as a reason for postponing such measures, taking into account that policies and measures to deal with climate change should be cost-effective so as to ensure global benefits at the lowest possible cost.[18]

However, the Third North Sea Conference (1990) specifies an even more permissive set of circumstances in which the precautionary principle can apply:

> The participants … will continue to apply the precautionary principle, that is to take action to avoid potentially damaging impacts of substances that are persistent, toxic, and liable to bioaccumulate even where there is *no scientific evidence* to prove a causal link between emissions and effects.[19]

Despite its appearance in documents such as these, the precautionary principle is not widespread as a policy making approach. One explanation – for which I lack evidence, but which is plausible – is that precautionary policy making is unlikely to be popular with the electorate, especially in straightened times and when precautionary measures are likely to be relatively expensive. This is especially the case in the wake of public health scares such as SARS, avian flu, and swine flu, none of which turned out to be as serious in terms of infection and death rates as experts initially predicted. And there may be good reasons to avoid blanket precautionary policy making. To adopt a precautionary approach to every single possible harmful outcome would not be a sensible use of limited resources – indeed, it might be inherently self-defeating. As Sunstein puts it, '[t]he regulation that the principle requires always gives rise to risks of its own … hence the principle bans what it simultaneously mandates'.[20] Applied in a blanket way at a general level, the precautionary principle is literally incoherent: each course of action it requires of policy makers it simultaneously prohibits.[21] And such a precautionary approach is certainly not attractive, if it is even possible, within a single life: '[t]he likely result would be paralysis, because so many courses of action would be forbidden. (Even doing nothing might be prohibited; human beings who do nothing … will probably end up pretty unhealthy. Will they even eat? What will they eat?)'.[22] In order to live and live well, most (sane) people ignore the possibility of, or take only very minimal precautions with respect to, harmful events below a probability threshold, which is a perfectly rational strategy.[23]

However, there is a deeper set of reasons related to scientific methodology which explain why the precautionary principle is rarely found in policy making. There are two broad types of states of knowledge to which scientists can lay claim: that a positive causal relationship exists between a process, event, activity, etc. and an effect, or that no such causal relationship exists. Classical scientific methodology is conservative insofar as scientists are more interested in avoiding false claims to the first type of state of knowledge ('false positives') than false claims to the second type of state ('false negatives'). The priority scientists give to avoiding false positives over avoiding false negatives can skew policy making dependent on expert scientific advice away from precautionary measures in regulation. However, the precautionary principle takes seriously the possibility of false negative knowledge claims – i.e. the possibility that claims that a given process etc. will not cause harm could be false – with respect to various processes etc. to which it might be applied. When policy makers insist on full scientific certainty – or something close to it – with respect to a process etc. *prior* to regulating that process, it is likely that the processes etc. for which the precautionary principle is designed (i.e. possible false negatives) will be neglected in regulation, because of the lesser attention given by scientists working on the relevant processes etc. to the possibility of false-negative knowledge-claims with respect to those processes etc.[24,25] Worse, as Harremoës et al. point out, even when attention is paid to this and a claim is made that there is 'no evidence of harm', this can be misinterpreted as 'evidence of no harm', and then used as the basis for rejecting precautionary measures even when calls for them are explicit, as was arguably the case in the BSE crisis in the UK in the late 1980s–1990s.[26]

Turning to the precautionary principle itself, it is sometimes presented as equivalent to, or derived from, the principle 'better safe than sorry'. So presented, it is hard to refute: who would rather be sorry than safe? However, these presentations are inaccurate to the precautionary principle as it appears in policy documents.[27] Rather, the precautionary principle removes constraints on reasons for action in policy making where scientific expertise is indispensable: it states that strong uncertainty about harm need, or must, not stand as a reason for inaction with respect to policy making that would adequately protect people from these harms. The principle has a weak and a strong formulation.

> *The weak precautionary principle*: When evidence or information is insufficient to establish the nature and/or probability of harms caused by an activity, policy makers are *permitted* to act in order adequately to protect people and other entities from these possible harms.

> *The strong precautionary principle*: When evidence or information is insufficient to establish the nature and/or probability of harms caused by an activity, policy makers are *required* to act in order adequately to protect people and other entities from these possible harms.

The weak precautionary principle is a permissive principle: it states the reasons on which it is acceptable for policy makers to act. The weak precautionary principle permits policy makers to take preventive action for reasons other than certainty that the action is necessary to prevent a harm. The strong precautionary principle is a categorical principle. It states the reasons upon which policy makers must act; that is, in order to protect people and other entities from possible harms, even when the nature and/or probability of these harms is not known. Given this uncertainty, reasons given in justification of action taken in the name of the strong precautionary principle must appeal to considerations other than those of evidence- or model-based support for the belief that the action is necessary to prevent a harm. It is relatively rare to find examples of the strong precautionary principle in policy literature; far more common are varieties of the weak precautionary principle.[28]

The weak precautionary principle is impotent as a constraint on the decisions of policy makers because it is satisfied even if decision making on every policy issue in its scope is postponed until the relevant evidence and information is certain; that is, even if no policy maker ever actually takes precautionary measures as set out in the weak precautionary principle. When one does not do P despite having permission to do P one acts consistently with that permission. Including the weak precautionary principle in policy documents need have no impact at all on any commitments appearing elsewhere in them, or in any other documents.[29]

For this reason I shall focus instead on the strong precautionary principle. This is a steely principle: it compels policy makers to act in the face of strong uncertainty even though their action may in fact turn out to be an unnecessary precaution against a non-existent, or vanishingly unlikely, harm. Because such action is almost always expensive (in literal sterling terms), and costly (in terms of the intrusiveness, disruptiveness, and unpopularity of legislation to satisfy the principle, and its impact on other policy commitments), the strong precautionary principle is generally dismissed as an unworkable and naïve proposal made by those in the grip of an ideology or suffering from distorted perceptions of risk.[30]

On the back of this rejection, it is sometimes then claimed that policy makers ought instead to employ the techniques of cost-benefit analysis in framing legislation to address strongly uncertain harms. The aim of performing a cost-benefit analysis of a public policy is to establish whether 'the aggregate of the gains that accrue to those made better off is greater than the aggregate of the losses to those made worse off by the policy'.[31] If this is the case, then the policy is economically efficient, and to be commended to that extent. Putting to one side the question of how losses and gains are to be understood, measured, compared, and aggregated, as well as the question of whether there are dimensions of value beyond efficiency to be brought to bear in the assessment of public policy,[32] the picture embedded in this move of presenting the strong precautionary principle and cost-benefit analysis as approaches to policy making which necessarily or always deliver opposed recommendations is somewhat misleading, for at least two reasons. First, even when cost-benefit analysis stands opposed to the strong precautionary principle, it nevertheless might be an approach that provides useful guidance as to which particular precautionary policy to adopt, once an overall precautionary approach has been decided upon for reasons not delivered by a cost-benefit analysis itself. It is certainly not the case that commitment to the strong precautionary principle *ipso facto* rules out the use of cost-benefit analysis in consideration of various implementations of the strong precautionary principle in policy. Indeed, some have claimed that any sane version of the precautionary principle is delivered by a cost-benefit analysis that '[places] a thumb on the cost side of the cost benefit analysis'.[33]

Second, and more strongly, it may be that cost-benefit analysis itself recommends a precautionary principle for a certain limited class of cases in which the possible harms in question are catastrophic. Although opposed to the precautionary principle as a general approach in policy making, Sunstein takes just this approach with respect to some specific cases; notably with respect to climate change. Despite his rejection of the strong precautionary principle as a general approach to policy making, on the grounds that '[t]he regulation that the principle requires always gives rise to risks of its own … hence the principle bans what it simultaneously mandates',[34] he argues elsewhere for a 'Catastrophic Harm precautionary principle', 'designed to provide guidance for dealing with extremely serious risks'.[35] This principle is heavily qualified and, as Sunstein happily admits, does not stand as a 'decision rule' for policy makers.[36] It guides policy makers facing uncertainty with respect to the worst-case scenario outcome of their decisions to weigh these scenarios against one another in terms of the losses involved in each scenario, and by considering whether precautionary action to avoid each scenario is 'extraordinarily burdensome'. If an uncertain outcome A would impose losses far greater than an uncertain outcome B, and if the costs of precautionary action to prevent A are not extraordinarily burdensome, then policy makers should take this action. However, if precautionary action to avoid A is very costly, and/or A is not much worse than B, then B ought to be addressed through precautionary action, unless the costs of this action are extraordinarily burdensome and/or there is another worst-case scenario C only slightly less bad than B … etc.[37]

As we shall see, and as Sunstein admits, his argument is tantamount to a limited endorsement of the maximin principle, commitment to which he attributes to Rawls.[38] To that extent his argument prima facie coincides with the first Rawlsian argument to be laid out in the next section. However, there are two points to note which should dispel this appearance. First, and following the later Rawls,[39] the injunction to make the worst-off as well off as possible embodied in the difference principle, and upon which the argument in the next section draws, is not made in terms of the maximin decision rule: Rawls' final statement of his reasoning for the difference principle avoids both the terminology of maximin, and the content of the rule itself. Instead, the reasons to which Rawls appeals are explicitly moral and political, as opposed to being derived from decision theory or rational choice theory. Second, Sunstein's Catastrophic Harm precautionary principle requires that worst-case scenarios be cardinally ranked in terms of their badness,[40] and his methods for assessing the losses created by outcomes by reference to which he suggests ranking be performed are the techniques of cost-benefit analysis. In particular, he defends the proposal that the value of a statistical life should be used in the assessment of costs and benefits, and that (with some qualifications), the value of a statistical life in different contexts can be worked out by ascertaining people's willingness to pay to avoid various risks to life and limb.[41]

Sunstein's defence of these cost-benefit analysis techniques is sophisticated and sensitive, and here is not the place for an extended discussion. I note it simply to indicate a further difference between his defence of a limited precautionary principle and those to be considered in the next two sections, neither of which makes use of or requires these techniques. Instead, the arguments articulate some implications of a commitment to deliver justice to future generations. In the next section I lay out a Rawlsian argument that makes this case, and I comment on its limitations. In the section 'Unbearable strains of commitment in Methane Nightmare' I develop a second (and new) Rawlsian argument which is not subject to these limitations, and which delivers a more potent justification of the strong precautionary principle with respect to climate change catastrophes.

The playing safe argument

Famously, Rawls offers an interpretation of equality whereby to treat people as equals means (among other things) to ensure a distribution of (dis)advantage among them that makes the worst-off group as well off as possible. In addition, Rawls conceives of justice as intergenerational in scope, governing relations across generations as well as within them. These commitments work to justify the strong precautionary principle with respect to climate change catastrophes as follows.

The strong precautionary principle is justified with respect to many climate change catastrophes because the worst consequences of not taking precautionary action are worse than the worst consequences of taking precautionary action, and choosing the former course of action is not consistent with treating present and future people as equals when we cannot assign a probability to each outcome, i.e. when we are strongly uncertain of each outcome, as is the case with respect to many climate change catastrophes. We can see this by adopting the Rawlsian conceit of the original position, and by imagining what parties choosing principles therein would say in justification of their choices to those whose interests they represent. What such persons could not do is take a bet on the probability of the worst consequences of no precautions being lower than the probability of the worst consequences of precautions, given the nature of these consequences. Call this the 'playing safe' argument; in order to make it in full I need to specify in more detail the nature of the consequences in question.[42]

The worst consequences of taking precautions against climate change catastrophes are that the costs of adapting to climate change turn out to be lower than the costs of taking precautions. The consequences of climate change may turn out to be nowhere near as dire as experts' worst fears, and it may be that tipping-points and their climate change catastrophes exist only in nightmares and Hollywood movies, or are extremely distant in time, giving us more opportunity for cheaper precautionary action at a later date. It may be that the Earth is far more insensitive to our activities than we think. In this scenario, we spend a lot of money and time – with all the associated opportunity costs this creates – in taking precautions against consequences now that are in fact not necessary. Estimates of how much it would cost to implement adequate precautions vary widely: for example, Bjørn Lomborg puts the figure at $37.632 trillion, whereas the European Commission (working with a generous stabilisation target of 550ppm to be met by 2100) chooses $1 trillion–$8 trillion,[43] and the *Stern Review* estimates the cost at 1 per cent of world GDP in 2050, or $1 trillion.[44] (Note for comparison that the UK government alone has – at the time of writing – spent £1 trillion on guarantees and bail-outs for banks in the wake of the financial crisis of 2007, which roughly converts to $1.5 trillion).[45] These figures are hotly disputed, but to make the case against precautions as powerful as possible I shall adopt the Lomborg estimate. Call this outcome 'Unnecessary Expenditure'.

The worst outcome of not taking precautions is that the worst predictions of the IPCC turn out to be very conservative,[46] that the gradualist paradigm is false, and that we are on the verge of various tipping-points, with everything this entails. The worst outcome of a decision not to take precautionary measures with respect to climate change is catastrophic à la Methane Nightmare.

With these outcomes identified, the playing safe argument for the strong precautionary principle with respect to climate change is prima facie powerful: because Methane Nightmare is so much worse than Unnecessary Expenditure, we ought not to act as if the worst-case scenario of taking precautions is more probable than this worst-case scenario of not taking precautions, and thus we ought to take precautions against climate change catastrophes. Acting as if the worst consequences of taking precautions are more probable than the worst consequences of not taking precautions is ruled out because the harms that would be caused to present and future people by policy made on this assumption, if it turns out to be false, are much greater than the harms that would be caused by policy made on the first assumption, if this turns out to be false. And treating people as equals makes it impermissible to gamble with their interests in this way.

Stephen Gardiner's elaboration (following Rawls) of the conditions under which Rawlsian reasoning delivers the requirement to take precautions is helpful.[47] These are when (a) decision makers are in a state of strong uncertainty with respect to the probability of the events in question; (b) decision makers are indifferent to gains above the minimum that would be guaranteed by a strategy to make the worst-off as well off as possible, i.e. their concern is focused solely on improving the position of the group with least; (c) the alternative strategies of distribution – wherein (b) above is not the priority – are unacceptable.[48] We have seen that (a) is true of many climate change catastrophes (I shall consider objections to this claim in the section 'Three problems'). Conditions (b) and (c) are questioned when it is claimed that it is rational or reasonable to gamble on the probability of Unnecessary Expenditure being higher than Methane Nightmare – and so not to take precautions – because if this gamble pays off the gains are so much higher than if the bet is not placed at all, or is placed on the probability of Methane Nightmare being higher than Unnecessary Expenditure.[49] Call this the 'sceptical stance'.

Two comments on the sceptical stance. First, it does not take seriously the possibility that even if the bet at its heart were to turn out to be good we may still have reason not to make it, in virtue of how taking precautions against climate change catastrophes could have all sorts of other benefits that would make this choice attractive even if any climate change catastrophe turns out to be highly improbable. This is the so-called 'no regrets' strategy: the requirement to take precautionary measures against climate change catastrophes could stimulate technological innovation, efficiency measures, and new forms of political interaction that could greatly enhance the quality of our lives even if the precautions turn out to have been redundant with respect to climate change catastrophes. In other words, the opportunity costs created by not taking precautions might make it always less costly to take precautions, even taking into account the possibility that the probability of any climate change catastrophe is tiny. Although the 'no regrets' strategy is not part of any Rawlsian argument and so, strictly speaking, not part of the playing safe argument, it is certainly not inconsistent with it. Indeed, it could be proposed as an interpretation of condition (c) above: not taking precautions is an unacceptable alternative not just because of the possibility of climate change catastrophes, but also because of the possible gains associated with precautionary measures. Of course, whether taking precautionary action against climate change is a 'no regrets' strategy is highly controversial. Hence, if there is an alternative, or complementary, way to undermine the sceptical stance, then all the better.

This brings me to my second comment, which relates to the type of justice to which the playing safe argument as it stands appeals. The argument is that climate change policy makers concerned to treat persons (remembering that this includes future generations) as equals should be guided to take precautions against climate change catastrophes by a concern to make the worst-off as well off as possible, because Methane Nightmare is much worse than Unnecessary Expenditure. In what ways, exactly, is Methane Nightmare worse than Unnecessary Expenditure? Clearly, there are all kinds of losses – in biodiversity, to ecosystems, of habitats – that would occur under Methane Nightmare but not under Unnecessary Expenditure. Let me put these to one side in order to focus on a harm people in Methane Nightmare would suffer that people in Unnecessary Expenditure would not. In Unnecessary Expenditure we waste resources that could have been used to improve the position of the worst-off in the current generation or, indeed, that could have been saved for the benefit of the worst-off in future generations. These are considerations of distributive justice, and in line with the impartiality at the heart of the Rawlsian approach to justice, duties of distributive justice are intergenerational in scope.[50] If we have a duty to ensure that the worst-off – understood in intergenerational terms – are as well off as possible, and if Unnecessary Expenditure conflicts with this duty, then we have a reason to reject courses of action leading to it, and should avoid it, all else being equal.

The playing safe argument as presented so far can be read as claiming that all else is not equal: the distributive injustice that would be created by Methane Nightmare is worse than that created by Unnecessary Expenditure, and we must choose between courses of action to which Unnecessary Expenditure and Methane Nightmare attach as worst outcomes, in which case we should aim to avoid the worst worse outcome – i.e. Methane Nightmare – by taking precautions. Any disadvantage that accrues to the worst-off as a result of this choice is justifiable to them in the name of equality: given the badness of Methane Nightmare, choosing not to take precautions – given our uncertainty – would be to gamble with their interests by betting on the probability of Methane Nightmare being lower than that of Unnecessary Expenditure.

However, as we have seen, making the argument in terms of distributive justice leaves it vulnerable to the sceptical stance, as follows. This version of the argument turns on the requirement to ensure that the worst-off generation is as well off as possible in terms of tangible goods, which includes generations in the future who may suffer huge disadvantage if we do not take precautions against climate change catastrophes. It can then be objected that tangible goods can be priced, and that the cost of taking precautions now will make these goods unaffordable for future generations. People who take the sceptical stance characteristically specify why it is rational or reasonable to gamble on the probability of Unnecessary Expenditure being higher than Methane Nightmare (and so not to take precautions) in terms of monetary gains and losses: crudely, we will be richer, and make future generations richer, if we do not take precautions. Of course, most people who take this tack also insist that monetary gains and losses matter because we can do things with money, such as improving human health.[51] However, as has been observed frequently, there are things we value (or ought to value) which cannot be compared in value to other valuable things, and so are not amenable to being priced. Consequently, such incommensurably valuable things do not appear in the calculations that inform rejection of precautionary approaches such as this.[52] One of these things is intergenerational justice.[53]

The playing safe argument as made so far is limited to a conception of intergenerational justice in distributive terms. However, there is another way to make the playing safe argument that articulates the impermissible gamble at its heart in terms of a more fundamental vision of what intergenerational justice requires, and evades the sceptical stance. I make that argument in the next section.

Unbearable strains of commitment in Methane Nightmare

There is something bad about Methane Nightmare that is related to justice, but is not well captured in terms of distributive justice, and which is absent in Unnecessary Expenditure. Highlighting this feature of Methane Nightmare makes available to us a further reason for taking precautions against climate change that justifies this choice not just in virtue of an egalitarian commitment not to gamble with the interests of the worst-off in accumulating as many tangible goods as possible, but also in virtue of a commitment to making it possible for future people to commit themselves to justice at all. This additional badness turns on the fact that Methane Nightmare would create unbearable strains of commitment for those living through and after it: it would make unreasonable any mutual expectations among future generations that they should propose and abide by fair principles of justice.[54] If this would be true of future generations in Methane Nightmare, then we – the current generation – cannot adopt principles for action on climate change that could have this outcome.

The strains of commitment are related to the idea of making an agreement in good faith; in Rawls' words, 'not only with the full intention to honour it but also with a reasonable conviction that one will be able to do so'.[55] Rawls claims that excessive strains of commitment cause people to withhold affirmation of principles of justice in two ways. First, people 'become sullen and resentful … ready as the occasion arises to take violent action in protest against [their] condition'; and second, people 'grow distant from political society … withdrawn and cynical [they] cannot affirm the principles of justice in … thought and conduct over a complete life'.[56]

Any attempt to implement principles of justice in Methane Nightmare – and many other climate change catastrophes – would probably have these effects on people. An indication of the temperature rises that might be experienced in Methane Nightmare is available by looking back at the PETM, to which it is thought that the collapse of methane clathrates greatly contributed.[57] In this period, average temperatures increased by 5°C–10°C.[58] One of the few models that address the effect of a global temperature rise of 5°C predicts severe desertification in already arid regions[59] – decimating most of the world's breadbaskets – combined with huge increases in precipitation at higher latitudes, making floods and storms more frequent.[60] The net result of a 5°C increase would be worldwide famine, severe conflict over water in the world's desertified regions, and extreme territorial insecurity for those living in the dwindling inhabitable areas as an archipelago of refuges is put under increasing pressure by hungry and desperate people.[61]

At a 6°C increase, however, things likely become very much worse. Mark Lynas suggests that the best reference point for this amount of warming is not the PETM, but rather the end of the Permian era 251 million years ago, when the 'Great Dying' extinction event, which wiped out 95 per cent of the planet's species, occurred as a result of a temperature increase of 6°C.[62] At the Permian-Triassic boundary, the oceans heated drastically and became devoid of oxygen, and so unable to support life, and also spawned superhurricanes which distributed heat to the poles (a further positive feedback).[63] Destabilised methane clathrates beneath the oceans released huge quantities of CH_4 which, if ignited, would have generated explosive, lethal blast waves travelling at 2km per second, and killing everything in their path.[64] The rotting carcasses of animals, and decaying vegetation, in the oceans released large amounts of hydrogen sulphide, obliterating any remaining life there,[65] killing most remaining land animals when released into the atmosphere,[66] and destroying the ozone layer. For temperature rises beyond 6°C, predicting the effects is guesswork, but looking to conditions on Venus gives some indication. It is crucial to note that the temperature increases that caused the Great Dying probably took 10,000 years to take effect, whereas – even on the conservative estimates of the IPCC – we could achieve such warming in 100 years. As Lynas puts it, '[i]f we had *wanted* to destroy as much of life on Earth as possible, there would have been no better way of doing it than to dig up and burn as much fossil hydrocarbon as we possibly could'.[67]

The extreme scarcity of resources and almost unimaginable conditions in Methane Nightmare would make the joint pursuit of justice impossible.[68] When there is not enough for each to have even the bare minimum for survival, to ask anyone to do anything other than pursue their own self-preservation and perhaps that of their family – such as to act according to the demands of justice – is to ask them to accept death so that others can live. In normal circumstances the sacrifices in self-interest required by justice are not unreasonable and, all else being equal, do not impose unbearable strains of commitment on those on whom they fall. But in circumstances as desperate as Methane Nightmare, where self-preservation dominates all other motivations, any justice-based request for self-sacrifice will at least endanger a person's chances of survival, if not actually require their death. What makes this request an unbearable strain of commitment is not just the content of the request in itself; rather, it is that this request be accepted as a requirement of justice, that is, as a measure against which those upon whom it falls can make no justified complaint.[69] If future generations know that precautions could have been taken by their predecessors that would have averted Methane Nightmare, and would have made the request unnecessary, then sullenness, resentment, alienation from principles from which the request is derived, and violent protest against the request, are the least we can expect from them.

If social cooperation is possible in such circumstances, it is not according to principles of justice. In Chapter 5 I analyse in more detail what it means to claim that justice is not possible in circumstances such as these, and I offer a justification for principles of triage as an alternative method for distributing scarce resources in these circumstances, where these principles are understood as keeping alive hope for a future return to social cooperation governed by principles of justice when (if) circumstances become more propitious.

For my purposes here, however, it suffices to note that the badness of Unnecessary Expenditure is not related to any unbearable strains of commitment it would create:[70] the putative reason for avoiding it is not that it destroys the possibility of justice, but that avoiding unnecessary expenditure allows for a more just distribution of goods by making more goods available for redistribution and/or saving now. That future generations are made less well off than they could have been because the current generation takes precautions against the climate change catastrophes that its activities could inflict on future generations is not obviously unjust (indeed, I have claimed that there are good reasons of justice for taking precisely this course of action). And that members of future generations know that they could have been made better off had the current generation not taken precautions does not create a strain of commitment for them that makes their joint pursuit of justice impossible. That future generations have fewer goods than they would have had, had precautions not been taken, is not in itself a reason for them to reject any requirements made of them by principles of justice in such reduced circumstances, especially given their knowledge that the decision of the current generation to save less in order to take precautions was made in conditions of strong uncertainty, and in the name of delivering justice to their descendants.

If not taking precautions against climate change catastrophes could impose conditions on future generations that would make the joint pursuit of justice by them impossible, then we – the current generation – cannot decide not to take such precautions. When thinking through our decisions from the point of view of justice, we are required to adopt the standpoint of impartiality, which in part requires that we should not choose principles of social cooperation that could impose on future generations conditions which we ourselves could not agree to. To choose not to take precautions against climate change catastrophes such as Methane Nightmare violates this fundamental requirement of impartiality.[71]

Three problems

The strong precautionary principle is a highly controversial principle, and the science of tipping-points and climate change catastrophes is young: bringing the two together delivers an approach open to many objections. I shall consider three of these here. The first addresses the subject matter of the strong precautionary principle; the second questions the fitness of the strong precautionary principle as a principle for policy makers in conditions of strong uncertainty, and the third addresses the scope of the strong precautionary principle.

I have claimed throughout that strong uncertainty attaches to many, but not all, climate change catastrophes. With advances in science it is likely that the set of strongly uncertain climate change catastrophes will diminish in size.[72] In that case, it might be objected, the events for which the strong precautionary principle is supposed to offer policy guidance may shrink to an insignificant number. Or, more strongly, it might be objected that there are no such events, i.e. that strong uncertainty does not exist, either per se, or with respect to any climate change catastrophes.

To address the strong version of the objection first, there are no experts who would claim that all climate change catastrophes are, at most, weakly uncertain. Even if all the climate change catastrophes we know to be possible are weakly uncertain, it remains possible that there are climate change catastrophes and associated positive feedbacks of which we are unaware. To deny this possibility would be foolhardy. Thus, those who make this claim probably rely on a more general form of the objection, viz. that strong uncertainty per se does not exist. Sunstein's discussion and dismissal of this version of the objection are instructive and persuasive.[73] In summary, he argues that even if it is true that people always subjectively assign a probability to various outcomes when making decisions, it does not follow that this subjective assignment reflects any objective facts about the probability of the outcomes, which may nonetheless be strongly uncertain. Of course, the philosophical debate in this area is technical and sophisticated, and the literature it has spawned is large. Suffice it to say here that in asserting the existence of strong uncertainty with respect to some outcomes I am aware that I stand opposed to Bayesians on this question – and am happy to do so.

With respect to the weaker version of the objection, viz. that some climate change catastrophes are strongly uncertain, and with the likelihood that this number will diminish, giving the strong precautionary principle an ever-decreasing number of events to which it applies, the following two things should be noted. Even if only one climate change catastrophe is strongly uncertain, the steps we ought to take to provide adequate precautions against this outcome – i.e. drastic reductions in emissions etc. – would by their very nature also serve as precautions against other climate change catastrophes. Precautionary measures against climate change catastrophes do not discriminate between them, because they all have their root cause in excessive greenhouse gas emissions. Measures to avert one climate change catastrophe stand as measures to avert all.

The second objection focuses on the strong precautionary principle's requirement that precautions must provide *adequate* protection against uncertain harms.[74] With respect to specifying what counts as adequate protection, there are two basic options:

(a) that we devote all of our resources and efforts to taking precautions;
(b) that we devote less than all of our resources and efforts to taking precautions.

Prima facie, (a) is implausibly demanding because it would leave no resources for the pursuit of other important social goals. However, (b) is wildly vague, and gives no useful guidance to policy makers. But being more precise about exactly how many, and what, resources are sufficient for adequate protection against a harm requires that we specify the probability of the harm: without this specification, the notion of what is adequate loses its anchor, and we must rely on guesswork. This fact about specifying what is adequate creates the following problem. Either we can specify the probability of harmful events to which the strong precautionary principle applies, or we cannot. If we cannot, then the strong precautionary principle provides no guidance with respect to making policy for those events. Given that we cannot specify the probability of climate change catastrophes, the strong precautionary principle provides no guidance to policy makers concerned with this aspect of climate change. There is a way of making the same objection that comes at it from the other end, so to speak, as follows. We must know that any given climate change catastrophe putatively governed by the strong precautionary principle is not bound to happen – i.e. that its probability is not 1 – otherwise we would not bother to think about taking precautions against it (they would be futile, given our certainty that it is inevitable) and would perhaps just throw a big party instead.[75] In that case, we are not in a state of strong uncertainty with respect to the climate change catastrophes that I have claimed fall within the scope of the strong precautionary principle. In sum, the problem can be posed as a dilemma: either our genuine strong uncertainty about climate change catastrophes makes the strong precautionary principle inappropriate as a principle for policy makers addressing them, or our application of the strong precautionary principle to climate change catastrophes shows that we are not in a state of strong uncertainty about them.

In response, I think it is possible to specify two limiting cases to our strong uncertainty about climate change catastrophes that do not undermine the strength and nature of this uncertainty.[76] These are that we know the probability of climate change catastrophes is not 0, and is not 1. That is, we know climate change catastrophes are not certain not to happen, and are not certain to happen, which means that we know that it is possible that precautions could make a difference, and that enacting them is not necessarily futile. Nevertheless, we remain strongly uncertain of the probability of these catastrophes within the range of anything more than 0, and anything less than 1. And given the catastrophic nature of the events, any probability in this range justifies precautions. This does not, however, mean that we are necessarily committed to devoting *all* our resources and efforts to precautions. If the strong precautionary principle is, sensibly, qualified with 'all else being equal', then it is possible to claim that all else is not, in fact, equal because there are other important and urgent demands of justice that require resources to be achieved: in the face of strong uncertainty, it is these demands that set constraints on what counts as adequate for protection, and not any further, more precise, specification of the probability of various climate change catastrophes. The question of what other demands of justice there are that are sufficiently important and urgent to play this role is, however, properly a political question, to be debated by members of the demos, and to be decided using appropriate procedures for democratic decision making. Although experts can provide valuable information necessary to preserve the quality of such decision making, the decisions themselves are ultimately up to us.

This relates to the third objection. There are various future events that are catastrophic to the same degree, uncertain in the same way, and would create similarly unbearable strains of commitment for future generations as climate change catastrophes, but which are not related to climate change and so would remain untouched by any precautionary policy adopted to mitigate climate change catastrophes; for example, comet collision. The Chicxulub crater in Mexico is the remnant of a 10 kilometre-wide asteroid that struck 65 million years ago, and is widely believed to have caused the extinction of the dinosaurs.[77] Given that such a collision would no doubt do the same to us, should we be developing missile programmes to defend ourselves against this possible future harm? If so, what about supervolcanic eruptions? Or invasions of malevolent aliens? Why should we devote our attention and limited resources to taking precautions against climate change catastrophes as opposed to other possible global catastrophes? Call this the 'priorities' problem.[78]

The priorities problem can be addressed by reflection on the following point. First, the climate change catastrophes to which the strong precautionary principle applies are anthropogenically caused, and this distinguishes them from comet collisions, supervolcanic explosions, etc. The fact that we are doing things that could be (quickly) moving us closer to various tipping-points gives us a reason for taking precautions against its catastrophic effects that we do not have with respect to other, non-anthropogenically caused catastrophes: that we are, or could be, causing the problem prima facie renders us liable for finding the solution, and gives us a reason to prioritise finding a solution to that problem over others we are not, or could not be, causing (such as comet collision).

The existence of this reason in part alleviates the priorities problem, but possibly varies in its potency across two cases marking either end of a spectrum. In the first case, our status as the cause of possible climate change catastrophes is bad option luck for us, the current generation: in the full knowledge that increasing greenhouse gas emissions could move us closer to tipping-points leading to climate change catastrophes, we are taking our chances and increasing our greenhouse gas emissions anyway. The correct analogy here is with the drunken driver who knowingly imbibes, and then knocks down and kills an innocent child: the driver is fully morally responsible, and legally liable, for the death. Just as the driver either ought not to drink, or if it is too late and she is drunk, ought to take a taxi, so we ought to enact precautions to protect future generations (the exact analogy for our circumstances vis-à-vis greenhouse gas emissions is probably the case of taking a taxi). Note that if the problem of climate change were not intergenerational in nature the drunken driver analogy would be incorrect. Instead, it would be open to us to claim that knowingly and intentionally increasing our greenhouse gas emissions creates risks of harm for no one other than us and that – invoking something like Mill's harm principle[79] – we are thus under no obligation to abstain from, or mitigate the effects of, our risky activities. An accurate analogy would instead be of someone with no dependents or other special obligations to others who chooses to take heroin, knowing the risks. But brute facts about the atmospheric life of CO_2, and the speed at which the climate change it causes occurs, mean that vulnerable future generations cannot be written out of the picture: they stand to us as the innocent child stands to the drunken driver.

The second case, lying at the other end of the spectrum, makes less headway with respect to the priorities problem. This is when being the cause of a (possible) harm carries no implications for blameworthiness and is nothing but a piece of bad brute luck for the agent. The analogy here is of a sober and careful driver who kills an innocent child through no fault of her own. That we think the agent's causal efficacy has *some* moral significance in cases such as these is shown by the fact that we would judge any such driver who experienced no regret at having killed a child – who simply shrugged it off as 'one of those things' – as seriously morally deficient.[80] However, explaining the significance of the agent's causal efficacy for moral responsibility, and (perhaps) legal liability, in such cases is far from straightforward.[81] If the current generation stands to future generations as the blameless driver stands to the innocent child she kills, then it is less clear what difference it makes to the priorities problem that the climate change catastrophes under consideration are anthropogenically caused in contrast to other catastrophes which are not. (Discussion in the next chapter will address this, and related, questions in more detail.)

However, in my view, and in the view of most experts, we are in fact closer to the 'bad option luck' than 'bad brute luck' end of the spectrum – if not, in fact, at it – in which case the priorities problem is more tractable than at first blush. But note that even if we were closer to – if not at – the 'bad brute luck' end of the spectrum, the priorities problem does not stymie the approach proposed in this chapter, once we attend to what the problem actually is. The additional work that the priorities problem asks for is required to justify *not* extending the strong precautionary principle to non-anthropogenically caused catastrophes. This lacuna in the account of the overall scope of the strong precautionary principle does not affect the argument that it extends *at least as far as* climate change catastrophes. Thus, we do not need to wait for these further refinements before starting to think seriously about how to formulate and implement precautionary policies for climate change catastrophes.

Nevertheless, it might reasonably be objected that the response just outlined at best only supports the prioritisation of anthropogenically caused catastrophes over those not anthropogenically caused: a priorities problem remains *within* the class of strongly uncertain possible anthropogenically caused catastrophes. For example, how should policy makers rank in their decision making the strongly uncertain but possible catastrophes of Methane Nightmare, global destruction by molecular nanotechnology (the so-called 'grey goo' scenario), large-scale nuclear disaster or nuclear war, and (further) global economic meltdown? What guidance is provided by the strong precautionary principle with respect to the priorities problem in this class of harmful events?

Sven Hansson suggests a way forward which provides the beginnings of a way to address the priorities problem in this form: specification of a threshold of probability above which possible harms must fall in order to be candidates for regulation governed by the strong precautionary principle. He rejects this approach on the grounds that '[t]here does not seem to be a clear and principled way to define a plausible cut-off level', stating that appeal to 'scientific reasonableness' as a way of identifying the threshold is ruled out on the grounds that it is 'imprecise and difficult to apply'.[82] However, there are actually two questions related to thresholds, and Hansson's dismissal of the thresholds approach is correct only with respect to one of these questions.

These questions are: (1) what is the probability threshold above which an anthropogenically caused catastrophic event must lie to be a candidate for precautionary regulation? (2) does any particular anthropogenically caused catastrophic event lie above the threshold?[83] (1) is properly a matter for purely political decision making: what we, as a society, deem to lie within the scope of our legislative and regulatory concern should be decided through democratic deliberation in public reason. With respect to this question, scientists have no special knowledge, and should not be consulted as experts. This avoids the danger (identified in Chapter 2) of 'administrative rationalism', described by Clive Spash as follows: 'the decision process is removed to the realm of the objective "scientific" manager who reveals the truth and conveys the "facts" as uncontestable commandments'.[84] It also echoes the European Environment Agency's call for 'stakeholder participation' in climate change policy making.[85] If we are a very risk-averse society, with a large surplus of resources, we may decide on a low threshold, which would likely multiply the number of catastrophic events against which we commit to taking precautions. If we are gamblers, or have very limited resources, we may settle on a high threshold and only take precautions against very likely catastrophic events.

Wherever the threshold is set, however, (2) must eventually be raised in our deliberations at the level of policy, and it is here that 'scientific reasonableness' is relevant to my purposes, although not in the way that Hansson envisages. Given that the catastrophes of concern here are strongly uncertain, it is not possible for scientists to assign probabilities to them, and so not possible for them to indicate whether any such event lies above the democratically decided threshold. The simple solution of setting a threshold and then asking experts whether various catastrophic events lie above it is not available. Besides which, even if such judgements could be made, it is quite possible that all such events would lie above the threshold, in which case the priorities problem would not have been addressed at all.

With a democratically decided threshold in place, the way forward with respect to the priorities problem is to ask experts to make pairwise comparisons between uncertain catastrophic events, and thereby rank them, by asking them the following questions:

(a) 'which event, E1 or E2, is more likely to lie above threshold T?', then
(b) 'what level of confidence do you have in this judgement?'

With respect to (a), it is important to note the following two points. First, we are not asking experts where to set the threshold of probability. That is, we are not asking them (c) 'which catastrophic outcomes are acceptable?' Rather, we are asking them a question that will help us to democratically deliberate about the question (d) 'which catastrophic outcomes should we prioritise?' The fact that the proposition 'all catastrophic outcomes are unacceptable' is consistent with the enquiry directed by (d) – i.e. even if we think that all catastrophic outcomes are unacceptable, we still need to know which of them to address first – shows that asking (d) does not entail asking (c). As I have claimed, (c) is a question for democratic deliberation.

Second, note that we are not asking experts (i) whether E1 or E2 lie above the threshold, nor (ii) whether E1 or E2 is likely to lie above it. Question (a) does not require any judgement at all as to the probability of E1 or E2, or the probability of E1 or E2 having a probability above the threshold. This means that there is no inconsistency between holding that E1 and E2 are strongly uncertain, and the expectation that experts will be able to answer question (a) about them. What the question requires is a judgement of the extent to which each event possesses the property 'more likely to lie above T than En'; and a judgement, say, that E1 is more likely than E2 to lie above T is consistent with the belief that the probability of E1, and the probability of its lying above T, is uncertain.

The uncertainty of the events about which question (a) will be asked makes it necessary also to ask (b) in order that the political community can make a considered judgement as to the weight that should be given to experts' answers to (a)-questions in democratic deliberation about setting priorities. Indication of the

TABLE 3.1 Qualitatively defined levels of confidence

Level of agreement or consensus →		
	High agreement Limited evidence	High agreement Much evidence
	Low agreement Limited evidence	Low agreement Much evidence
	Amount of evidence (theory, observations, models) →	

Source: Adapted from IPCC, 'Guidance Notes for Lead Authors of the IPCC Fourth Assessment Report on Addressing Uncertainties', 2005, Table 2, p. 3.

TABLE 3.2 Quantitatively calibrated levels of confidence

Terminology	*Degree of confidence in being correct*
Very high confidence	At least 9 out of 10 chance of being correct
High confidence	About 8 out of 10 chance
Medium confidence	About 5 out of 10 chance
Low confidence	About 2 out of 10 chance
Very low confidence	Less than 1 out of 10 chance

Source: IPCC, 'Guidance Notes for Lead Authors of the IPCC Fourth Assessment Report on Addressing Uncertainties', Table 3, p. 3.

level of confidence in a judgement is an existing IPCC practice. The IPCC states that '[a] level of confidence … can be used to characterize uncertainty that is based on expert judgement as to the correctness of a model, an analysis or a statement'.[86] Its 'Guidance Notes for Lead Authors of the IPCC Fourth Assessment Report on Addressing Uncertainties' outlines a qualitative and a quantitative way of representing levels of confidence.[87]

The IPCC recommends that with respect to judgements over which there is high agreement and for which there is much evidence, the quantitative as well as qualitative classifications should be used (note that it would nevertheless be possible for such a judgement to have a low or very low level of confidence).[88] Although the IPCC recognises that its practice in the Third Assessment Report created some confusion over the relationship between levels of confidence and probabilities,[89] it is important to note that levels of confidence in a judgement do not correlate with any probability assigned to an event etc. by that judgement; for example, a judgement with a very high level of confidence may assign a very low probability to an event.[90] What level-of-confidence judgements add to the process of expert consultation in democratic decision making is information that enables us to justify giving more or less weight to expert judgements in our deliberations. Of course, it is always open to us to simply ignore all answers to (a) and (b) questions. But it is most unlikely that sensible democratic discussion would counsel such cavalier disregard for expert opinion.

This approach has some obvious limits, and raises questions, two of which I shall mention here. First, it will only narrow, and not eliminate, the range of hard cases created by problems of priority setting with respect to anthropogenically caused catastrophes. Those judgements in which experts have very low or low confidence are unlikely to be given weight in democratic deliberation about such policy. With respect to this remainder of hard cases there is nothing for it but to judge as best we can, using standards informed by the values we cherish, and accept the possibility of error. However, in the absence of crystal balls delivering a perfect ability to predict the future, the charge applies to all policy making – indeed, to all – decisions: there is always a possibility that an action performed with a purpose P will result in ~P. Given that the possibility of backfiring is a fact of the human condition, aiming to magic it away in the name of greater coherence in policy making is silly.

Second, there is an important question about how it gets decided which uncertain catastrophic events are subjected to the scrutiny of experts. Although we should adopt whatever procedures are necessary for the job, policy making on these matters would grind to a halt if the job were to include scrutiny of any and every possible catastrophe suggested as a candidate. This is, again, a matter for democratic deliberation in public reason, rather than expert judgement.[91] Take the catastrophe of 'God's Retribution': unless we mend our evil ways God will visit furious anger upon us with fire and brimstone. As we do not know the mind of God, this outcome is strongly uncertain. And if it happens, we will have been the cause of it. If religious groups can make a case in public reason – i.e. a case which could be accepted by free and equal people who may differ from them in religious beliefs, or have none at all – that we as a society ought to take the possibility of God's Retribution seriously enough to subject it to the scrutiny of experts in the ways just outlined, then so be it. In fact, such cases are unlikely to succeed, given that they rely on a faith particular to the religious practitioners in question, and not on values on which there is a reasonable overlapping consensus. But the structure and aims of the consultation process do not in themselves automatically rule out consideration of such events. It is very important for stability, and democracy, that religious etc. groups are not *ipso facto* excluded from these processes even if, in fact, they are very unlikely to succeed in getting policy makers and experts to consider their Armageddons.

At this point, religious believers may present themselves as experts in competition with scientists etc. in order to stake a claim to participation in the consultation procedure, or to the institution of a separate procedure for the scrutiny of their concerns.[92] However, religious believers can present themselves as experts if, and only if, the matters on which they pronounce lie beyond the reason of policy makers and, by extension, citizens. I submit that the existence of an angry God is not such a judgement. In general, the matters addressed by religion – good and evil, the meaning of life, the nature of salvation – are not beyond the ken of citizens and policy makers. Religious believers are not experts on anything beyond their own doctrine, holy texts, rituals, form of worship, etc. And even then religious belief is not a necessary condition for such expertise (one need not be a believer to be a theologian, or an anthropologist).

Conclusion

In conclusion, I have distinguished between two Rawlsian arguments for the strong precautionary principle with respect to climate change catastrophes. Although both are persuasive, ultimately the 'unbearable strains of commitment' argument provides the most powerful categorical grounds for taking precautionary action against climate change catastrophes. Overall, I have argued that the nature of climate change catastrophes requires us to take drastic precautions against further climate change that could lead us to pass the tipping-points that cause them. This is the case notwithstanding the fact that we are in a state of strong uncertainty with respect to these events; indeed, our strong uncertainty with respect to them – given their nature – makes the case for action to prevent them even more persuasive, from the point of view of justice. Some people translate the strong uncertainty of climate change catastrophes into weak uncertainty in order to justify taking precautionary action using risk assessment.[93] My argument is complementary to theirs. If divergent approaches to the uncertainty of climate change catastrophes nevertheless converge on the precautionary principle, then we have what Cass Sunstein calls an 'incompletely theorised agreement' on a policy, i.e. an agreement to which parties divided by often deep theoretical differences can nevertheless give their assent,[94] which is all to the good from a political point of view. In the specific case of climate change and its possible catastrophes, the fact that such an agreement has emerged provides a sliver of hope in the face of the increasingly dismal prospects for the planet uncovered by climate change science as it progresses. At the level of principle, we are agreed, whatever justificatory reasons we advance. What is needed now – in the immediate present – is policy to promote our principles, and the political will to enact it.

4

INTERGENERATIONAL CORRECTIVE JUSTICE

Introduction

As we have seen, climate change presents problems of justice in global and intergenerational arenas that are unprecedented. The seriousness of the harms it creates, the speed at which it is happening, the changes (to, for example, energy regimes, economic organisation, and consumer mentality) required to make any impact on it, and the frightening uncertainties that remain with respect to climate 'tipping-points' – beyond which presently unknown or poorly understood positive feedbacks could intensify and cause runaway climate change – show all existing policy approaches to be inadequate for prevention, mitigation, and adaptation.

This chapter offers the beginnings of a partial remedy to this inadequacy by articulating a proposal that reflects what any current generation owes to any future generation in the name of justice, in virtue of its activity causing climate change that puts those future generations at risk of harm. The proposal could be supported by various conceptions of justice framed in corrective, distributive, and retributive terms. My primary focus will be on how the perspective of corrective justice works to support the proposal, but I shall also comment on other perspectives of justice along the way, especially that of distributive justice.[1] At this point, however, it should be noted that these perspectives are not mutually exclusive, and may converge in the support they offer so as to create an overlapping consensus which is all the better for the proposal, philosophically and politically. As noted previously, Cass Sunstein calls this approach argument in terms of 'incompletely theorized agreements … not on high level theories about what is right or what is good, but on practices and low level principles on which diverse people can converge'. The more of these synergies we find, the more confident we can be in advocating the principled approaches to climate change policy on which they converge. The principal aim of this chapter is to articulate the conception of corrective justice that has a role in any incompletely theorised reasonable agreement on intergenerational duties of justice related to climate change.

In the previous chapter I argued for a precautionary approach to climate change policy by focusing on the possibility that our current greenhouse gas-emitting activities may be the cause of catastrophic climate change as a result of which future generations could suffer serious harms. However, even if such an approach were to be adopted by climate change policy makers – which is, at best, currently no more than a hope – the nature of the uncertainty attaching to climate change catastrophes means that any precautions taken may turn out to be inadequate. It may be that emissions are already sufficient to cause future climate change catastrophes, and that in virtue of passing key tipping-points of which we are unaware, such climate change catastrophes are already in the pipeline. But even if catastrophe is not on the cards for future generations, it is nevertheless highly probable that they – at least the six to eight generations who will follow us – will suffer the effects of our emitting activities.[2] For both these reasons, I shall argue, we ought to save now in order to provide future generations with funds and resources adequate to compensate them for the harm at which we are possibly (with respect to climate change catastrophes) and probably (with respect to non-catastrophic but still harmful climate change) putting them at risk.

As we have seen, one influential approach to intergenerational justice is Rawls' just savings principle. This requires the current generation to save for the benefit of their successors at a rate sensitive to the current generation's 'stage of civilisation', roughly, their level of affluence, and the extent to which they have achieved – or could achieve, given their affluence – just institutions: the just savings principle mandates different rates of saving for differentially affluent generations. In this chapter I shall offer a justification of a just savings principle that shows its suitability as a principle to address the potential harms to future generations of our current activity causing climate change, and I shall draw upon considerations of corrective justice in making this case. At various points I will call upon Rawlsian arguments and values in making the arguments. However, to my knowledge, Rawls nowhere discusses, or even mentions, corrective justice.[3] Nevertheless, in this chapter I shall elaborate a version of his just savings principle as a principle of intergenerational corrective justice, and shall defend it from within a broadly Rawlsian normative framework.

The arguments I shall make here relate to intergenerational justice, and so are of consequence for many issues beyond climate change, because there are many ways in which we can harm those who will follow us. However, they are of particular significance for climate change, given the gap between advocates of distributive justice committed to using all resources available now exclusively to benefit the world's current (and, at most, next-generation) disadvantaged,[4] and those committed to saving some of these resources for future strangers.

Disregarding the fact that this position rests on a false dichotomy – either we can do justice to current people, or to future people (why not both?) – an account of a just savings principle qua principle of corrective justice challenges the central claim in this position. Even if it is true that we cannot deliver distributive justice to both current and future people, and that we should prioritise delivering distributive justice to current people, we may nevertheless also have duties of corrective justice to future people that require us to save for their benefit.

That corrective justice provides a useful perspective on some of the political problems of climate change is not in itself a new idea.[5] What is new here, however, is the suggestion that corrective justice has an important place in our thinking about issues of *intergenerational justice* raised by climate change once we focus on the *risks* that the current generation impose on future generations by their greenhouse gas-emitting activities. I shall begin by commenting in general terms on the idea of corrective justice.

Corrective justice

There is more to justice than the requirement that goods be distributed (or not) according to a given (or no) pattern. In addition (at least), justice consists in ensuring that any party violating the rights of another, and in that way causing them harm, make reparation (often in the form of compensation) to the victim so as to redress the imbalance of justice between them. This is the subject matter of 'corrective' justice. Principles of corrective justice are principles of transactional justice: they address the question of who owes what (if anything) to whom when transactions between parties do not conform to standards independent of the principles of corrective justice themselves.

Jules Coleman correctly identifies the 'core' of duties of corrective justice in the following terms. A person A suffering harm has a claim against a (putatively liable) agent B under corrective justice if and only if the following conditions hold:

(1) The importance of human agency: the harm to A is a result of human agency (as opposed to, for example, the forces of nature);
(2) Correlativity: the harm to A is connected in a normatively significant way to B's actions (perhaps B caused the harm through her agency, or is otherwise responsible for it);
(3) Rectification: the appropriate form of A's claim against B takes the form of a claim for repair or rectification (often understood in practice as a claim for compensation).[6]

These three conditions are necessary and jointly sufficient for B to be liable to A for compensation only if B has an independently specified duty of care (which could be derived from a duty of justice) to A. They make it possible to distinguish duties of corrective justice from, for example, duties of benevolence, or duties of retribution. In particular, they make it possible to distinguish duties of corrective justice from duties of distributive justice. Relations of corrective justice are unlike relations of distributive justice insofar as they connect a particular party with another just in virtue of an action performed by the liable party. Corrective justice reflects particular features of the relationship between parties that may not (very likely – hopefully – will not) be true of the parties' relations with all other members of their society, i.e. liability (on the part of the injurer) and suffering (on the part of the victim of harm). In contrast, distributive justice structures the relationship of each person in society with all others regardless of how each acts or affects the others: it determines her place in a distribution of goods judged to be just in virtue of how each position in that distribution compares with all the others in terms of the goods allocated to it.[7]

I aim to show that many of the harms of climate change that the current generation could inflict on future generations are sufficient to render the current generation liable to any future generation it puts at risk of these harms, and that any future generation for whom these risks materialise has a claim against the current generation for compensation that correlates with the current generation's liability. It is highly plausible to add to this that any current generation has a duty to provide funds adequate to the just claims for compensation that many of their successors will have against them. The current generation should discharge the obligations created by this liability through contribution to an intergenerational fund from which any future generation suffering harm can draw adequate compensation.

The idea of corrective justice per se, insofar as it underpins claims that victims of harm have against their injurers, has considerable intuitive force: if my upstairs neighbour floods my home through her carelessness in leaving her bath running, we think that I have a just claim for compensation against her. The intuitive force of corrective justice at the interpersonal level is retained, I think, when the proposal is extended in scope to international contexts: if the small island state of Kiribati (made up of thirty-three low-lying atolls in the Pacific) disappears as a result of global warming caused by the climate-changing activities of US gas guzzlers and Chinese coal burners, then Kiribati has a just claim for compensation against the USA and China.

However, an important dissimilarity between the intergenerational, and the intragenerational and international, contexts stretches the intuitive appeal of corrective justice with respect to relations between generations: in the latter victims and harmers are contemporaneous, whereas in the former most are not. With respect to the intergenerational context, putative victims and harmers exist at different points in time: putative harmers exist before their actions have created victims (indeed, often before anyone yet exists who could be made a victim).

However, this does not mean that corrective justice has no role to play in any incompletely theorised agreement on principles of intergenerational justice related to climate change, because the conception of liability appropriate to corrective justice can be modified so as to make the non-contemporaneity of harmers and victims in the intergenerational context non-essential to their status as harmers and victims; and there are good independent reasons for making this modification. In the next section I shall argue that this modification is attractive insofar as it eradicates luck from the allocation of responsibility for losses effected by principles of corrective justice. Furthermore, I shall argue that the modification is required from the point of view of a commitment to equality, understood in Rawlsian terms as ensuring that those worst off in the context of corrective justice are made as well off as possible.

The ensuing discussion – in the two sections subsequent to this one – will take us some distance from consideration of problems of climate change as they currently exist. However, without the theoretical grounding provided in these sections the final proposal – that an Intergenerational Climate Change Compensation Fund should be established in the name of intergenerational corrective justice – will stand on sand.

Ex ante responsibility for risk-imposition

Many contemporary legal scholars analyse the law of torts in terms of corrective justice.[8] As Ripstein puts it, '[t]ort law answers two of the most fundamental questions faced by any society: "how should people treat each other?" and "whose problem is it when things go wrong?"'.[9] The corrective justice approach has it that tort law answers these questions by using liability rules to assess the putative wrong done to a victim by an injurer's actions in terms of violation of the law in general, where scholars in this school see the law as expressing moral norms of behaviour informed by reciprocity,[10] conventional rules of reparation for the violation of rights,[11] or the equal rights of persons understood abstractly as purposive beings,[12] to name three prominent alternatives. I shall not enter this debate here, and will merely note the following. In contrast to these theorists, I do not bind myself to offering an account of corrective justice that can make full sense of existing tort law (if any theoretical approach can): indeed, the account I will offer of risk-imposition as grounds for liability is anathema to most theorists of tort law. My discussion is instead pitched at a higher level of abstraction through a focus on (a) what principles of corrective justice should be adopted, given an impartial concern to treat people as equals; (b) how such principles affect our thinking about relations of intergenerational corrective justice; and (c) what this means for the delivery of climate change justice to future generations. That said, I will make use of a particular interpretation of corrective justice by a theorist of tort law, Christopher Schroeder;[13] in particular, of some of his proposals for adjusting tort liability.

Schroeder conceives of the aspirations of corrective justice as 'to ground legal liability on norms of moral responsibility and autonomy, and to distinguish clearly between these norms and other "external objectives" that law might serve, such as distributive justice'.[14] He argues that any tort system must meet three conditions, each of which is necessary, and which are jointly sufficient, for it to qualify as an expression of corrective justice so conceived. When a victim has a claim for compensation against a putatively liable injurer:

> (1) individual liability must be assessed consistently with moral norms of responsibility for one's actions ['action-based responsibility'];[15] (2) victims must be made whole (compensated) ['just compensation']; and (3) the resources for satisfying (2) must come exclusively from the liability payments required by (1) ['internal financing of compensation'].[16]

Let me focus on (1), which mirrors Coleman's 'correlativity' requirement for an injurer to have a duty of corrective justice to a victim (see p. 74). A key part of the way in which tort law has often been interpreted as satisfying (1) is through the requirement of cause-in-fact:[17] a party is liable under tort law if and only if the harm suffered by a victim was caused by them. That the defendant put the victim at risk of harm – even if the defendant knew that her action was risky – is insufficient for liability: harm has not been suffered by a victim and, *ipso facto*, cause-in-fact is absent. However, Schroeder argues that requiring cause-in-fact is not the only way in which liability can be interpreted so as to realise corrective justice. Indeed, there is an interpretation of liability available that better satisfies the requirement of action-based responsibility, and makes cause-in-fact entirely redundant by focusing instead on risk-imposition.

Making cause-in-fact necessary for liability imports what Schroeder calls an *ex post* interpretation of responsibility which,

> [P]ermits appraisals of responsibility to turn on facts or theories that could not have been known to the actor prior to her taking the relevant action … From this viewpoint, it is eminently possible for Jones and Smith to perform otherwise identical acts, but for one to escape any financial responsibility whatsoever, while the other is held liable for massive damages.[18]

According to Schroeder, the *ex post* view of responsibility injects an unacceptable degree of moral arbitrariness into assessments of liability: two persons who perform identical acts in identical circumstances suffer very different consequences as a result of arbitrary features of the world which make a difference to the effects of their actions on others subsequent to the performance of the action.[19] It is on the back of a modification of the *ex post* view of responsibility in response to this problem that I will develop an account of intergenerational liability for the imposition of risks of climate change harm.

Jeremy Waldron makes the problem of arbitrariness vivid by means of an example involving two equally careless drivers, Fate and Fortune. Fate's carelessness (in taking his eyes off the road for a few seconds) causes him to hit a motorcyclist, Hurt, and inflict severe injuries. Fortune behaves in an identically careless way, but causes no injury. Fate faces a $5 million law suit as a result of his act of carelessness, whereas Fortune faces no penalty at all.[20] Waldron assesses the arbitrariness of tort law's asymmetrical treatment of Fate and Fortune thus:

> Hurt's disability will always be something that Fate was *responsible for*, always something that he will have to regard as part of his history, always something marked, so to speak, on his account. Nothing like this can be said about Fortune. Unhappily for the advocates of tort liability, however, the connection between causation and responsibility fails to establish a non-arbitrary basis for differential liability as between Fate and Fortune ... On one side, it means little more than 'caused' (the avalanche was responsible for the destruction of the village). On the other side, it connotes culpability and liability. It is not in dispute that Fate's carelessness *was* responsible (and Fortune's carelessness was not responsible) for Hurt's injury in the first sense. But something more is needed to establish the relevance of the second sense of responsibility in a meaningful way.[21]

Schroeder advocates an *ex ante* conception of responsibility that addresses the problem of arbitrariness. On the *ex ante* view, Schroeder claims, an agent's responsibility is to be assessed by considering the extent to which she could have known, should have known, or did know that the action she performed would harm another;[22] whether she should or could have acted otherwise; and whether her agency was genuine (whether the action performed was authentically hers). On this view, it is consideration of the agent *up to and at the time of acting* that matters for responsibility, not what happens after she has acted.[23]

It is important to note that according to an *ex ante* conception of responsibility, cause-in-fact is still relevant to corrective justice (and tort law, to the extent that it expresses it) because it is required in order to identify which persons have a claim for compensation, i.e. those that have actually been caused harm by the actions of others. Shifting to an *ex ante* approach leaves intact this aspect of the relationship of corrective justice between harmer and victim. Instead, the shift marks a different interpretation of the action-based responsibility condition for tort liability, whereby liability must respond to what the putative harmer does. With respect to this condition, an *ex ante* approach limits attention to what is true of the agent's actions (and accompanying intentions etc.) up to and at the point of acting. As Schroeder puts it, '[i]t is the connection of causation to the defendant that an ex ante theory of tort severs, not the connection of causation to the plaintiff'.[24,25]

Whether or not Schroeder and Waldron are right about tort liability, we can extrapolate from their claims to make sense of what people owe to one another as a matter of corrective justice. The key point for my purposes is that it can render liable agents who perform acts putting others at risk before the risk materialises, or even if the risk never materialises, and even when those others are unaware of the risks that have been imposed on them.[26] Prima facie, this means that any current generation putting future generations at risk of climate change harms lies within the scope of *ex ante* liability at the point at which, for example, they emit greenhouse gases. Incorporating the *ex ante* view of responsibility into an account of corrective justice allows it to meaningfully stretch out in time by providing a justification for extracting resources for reparation from a liable agent (or her insurers) now, even though those their actions put at risk exist in the distant future and, indeed, even if many of those distant strangers are never in fact harmed by the action. This justification is in place if it can be established that the agent's action breaches a duty of care she has to the victim not to impose such a risk on her, that the action could have been avoided by the agent, that the agent is responsible for imposing the risk through her action (perhaps, in part, by causing it to be imposed), and that she is the authentic agent of the action (she has not been brainwashed, drugged, etc.).

The force of the shift to an *ex ante* approach can be seen by considering what difference it might make in practice. Let me first take Schroeder's example of the Speeding Motorist:

> A motorist, in a hurry to get from her house to an important engagement in the neighboring town, enters the connecting freeway and speeds up to ten miles per hour over the posted speed limit. She arrives at her destination safely, only a few minutes late.[27]

On an *ex post* view of responsibility, the Speeding Motorist, having caused no harm, is not liable under corrective justice. What difference does it make to liability once we consider the case from the point of view of corrective justice using an *ex ante* conception of responsibility? Schroeder suggests the following:

> Imagine that an onboard computer registers the fact that Motorist is speeding and immediately authorises a withdrawal from Motorist's checking account in an amount equal to the value of the estimated increase in driving related accidents attributable to motorists speeding ten miles per hour above the speed limit. The money goes into a fund accessible by individuals injured by speeding motorists. Again, liability is complete at the moment of speeding. It does not increase if Motorist actually causes an injury, and does not decrease if Motorist does not cause an injury.[28]

This is recognisably a proposal to realise corrective justice because, to recall Coleman's characterisation, it responds to harms caused by human agency (driving-related accidents), connects the victims of these harms with the agents of them (those whose risky behaviour causes accidents) in a normatively significant way (via their *ex ante* responsibility for driving-related accidents), and enables the injustice against the victims of these harms to be rectified (through compensation for harm caused by driving accidents drawn from the fund created by fining all motorists who drive over the speeding limit). Treating Speeding Motorists in this way meets the requirements of corrective justice just as well as an *ex post* approach; '[t]he difference is that it requires compensation from all participants in a class of actions, rather than only from those whose actions fortuitously cause ultimate harm'.[29] Coleman makes the point as follows (without defending the proposal):

> One might object to luck playing such an important role in allocating accident costs. The problem can be solved by paying heed to the distinction between the grounds of recovery and liability. Suppose we accept as our principles of recovery that individuals who suffer harm at the hands of the faulty are entitled to recover, then instead of imposing the costs of particular accidents on respective faulty injurers, we might distribute the total costs of accidents among all agents who are at fault, whether or not their fault occasions harm to anyone.[30]

One obvious objection to the shift from *ex post* to *ex ante* can be made by reference to Fate and Fortune. It could be claimed that there is a simple way to mark the difference between: Fate actually inflicts, and Fortune does not inflict, injury on Hurt, and this is why they ought to be held differentially liable, and why the *ex ante* approach is to be rejected.[31] Of course, this difference between Fate and Fortune cannot be denied. But it is tantamount to no more than the fact that Fate caused injury to Hurt, whereas Fortune did not, and it is the normative significance of precisely this fact to establishing Fate's liability for the injury that requires unpacking (see the quote from Waldron above).[32] Nevertheless, consideration of this objection creates an opportunity to make it explicit that the idea of responsibility as it matters for corrective justice is *normatively loaded from the start*, and does not reflect a set of metaphysical facts (if such there are) about who is *really* responsible for an act and its consequences. As Ripstein argues, the idea of responsibility as it affects the terms of our social cooperation (some of which ought to embody corrective justice) is socially constructed: it captures what we think matters to how people ought to treat one another in socio-political contexts. There may well be a metaphysical idea of responsibility at the heart of different debates about morality, or free will, but it does not (and should not, given the burdens of judgement and the reasonable pluralism on such matters that they make inevitable and not-to-be-regretted)[33] bear on this political debate.

So far I have shown that it is possible to adopt a conception of liability (i.e. *ex ante*) that responds to risk-imposition, and so carves out space for intergenerational liability for the imposition of risks of climate change harm, under corrective justice. Now, as promised earlier, I shall show that we have independent reasons to adopt an *ex ante* approach.

I shall start with a question: which conception of responsibility ought to be adopted, given a commitment to equality and impartiality in the construction and operation of principles of corrective justice? I submit that this commitment shows the *ex ante* conception to be preferable to the ex *post* conception of corrective justice along two dimensions: (a) with respect to the interests of persons as potential victims of harm, and (b) with respect to the interests of persons as potential agents of harm.[34]

(a) The *ex ante* system of liability is preferable to the *ex post* system with respect to persons' victim-interests because it ensures in advance of any actual infliction of harm that a fund is available from which adequate compensation can be drawn for victims, whereas the *ex post* system cannot guarantee the availability of compensation (perpetrators may go AWOL, be insolvent, or be 'judgment-proof').[35] Thus, the worst-off victim under an *ex ante* system will be better off than the worst-off victim under an *ex post* system; and given that the *ex ante* system satisfies the correlativity requirement of corrective justice just as well as the *ex post* approach, equality – insofar as it involves ensuring that the worst-off are made as well off as possible – entails *ex ante* principles of corrective justice.

(b) The *ex ante* system is preferable to the *ex post* system with respect to persons qua potential agents of harm because an *ex post* system makes some agents carry the full costs of compensating victims, despite the fact that there is no morally relevant difference between them and others who performed actions creating identical risks, but who were lucky enough not to cause any actual harm (the arbitrariness problem). The *ex post* system makes a morally arbitrary distinction between equally culpable persons through a focus on whether their actions in fact caused harm, and this distinction confers on persons the (justice-related) disadvantage of prosecution and liability for compensation (realised in tort law or otherwise), or the advantage of freedom from it. Because the disadvantage of the worst-off perpetrator under the *ex post* system is not consistent with a commitment to equality so understood, and because the *ex ante* system eliminates the possibility of this morally arbitrary disadvantage, the *ex ante* system is preferable to the *ex post* system.[36]

Let me refer back again to Fate and Fortune in order to clarify the nature of the bad brute luck in question here, the effect of which is, I am arguing, eradicated by shifting to an *ex ante* conception of liability. Fate and Fortune are equally at fault, and equally responsible, for their careless driving: if they are caught by the cops and prosecuted for this no injustice is done. The prosecution would be bad luck for the both of them, and they deserve what they get. This is consistent with – indeed, entailed by – the *ex ante* approach. What differentiates Fate from Fortune is not his careless driving, but the fact that a motorcyclist happened to cross his path at the precise moment of his carelessness. His bad brute luck is this coincidence, for which he cannot be held responsible. The disadvantage it visits upon him – being a result of this brute luck – is not justified from the point of view of equality, if we think that equality involves correcting for brute-luck inequalities.[37]

How might an *ex ante* system of corrective justice governing relations between generations be modelled in practice? There are at least two sets of possibilities. First, intergenerational litigation, whereby principles of tort law conceived in terms of corrective justice are appealed to by future generations in bringing suits against past generations for redress for harms suffered as a result of risks imposed on them by past generations. Applied to climate change, this approach mandates legal action by members of future generations against members of past generations for climate change harm suffered by members of future generations caused by the greenhouse gas emissions of members of generations that preceded them. Or second, the requirement that any generation imposing a risk of harm on any subsequent generation should make an appropriate contribution to an intergenerational fund against which any members of a future generation may make a claim for reparation in the name of corrective justice, if the risk ripens into a harm to them. Applied to climate change, the requirement is that an intergenerational climate change compensation fund be established to which future generations may apply if harmed by the emissions of past generations, and to which members of the current generation must contribute, all else being equal. I favour this option for realising intergenerational corrective justice with respect to climate change, and shall elaborate on the reasons for this in due course.

My defence of the corrective justice approach to aspects of intergenerational justice means that such a fund can usefully be viewed as the realisation of Rawls' just savings principle, requiring a positive rate of savings for any generation whose external circumstances are sufficient to support just institutions, independent of any requirements of distributive justice they also stand under. Because the *ex ante* system of corrective justice responds to risks of harm caused by human agency, it is possible that some generations have no obligation of corrective justice to save, because it is possible that some generations impose no risk of harm on members of any future generations. The corrective justice interpretation of a just savings principle does not necessarily translate into a blanket requirement to save at a positive rate for every generation in history. Generations prior to the Industrial Revolution likely breached fewer (if any) duties to the current generation through the imposition of risk than generations subsequent to the Industrial Revolution. In non-industrial circumstances, the limited technologies and causal influence on the world of such generations would have made it possible for them, at most, to breach their duties by imposing risks on just one or two generations subsequent to them. However, it is very improbable that any generation in a developed and industrialised context imposes no impermissible risks on generations subsequent to it; and this is certainly not true of the current generation.

Before moving on, consider a potential problem with the approach as developed so far by way of clarifying the nature of *ex ante* liability still further. Risks are, by their nature, uncertain, and as such may never materialise. If that happens, will an injustice have been done to the generation from which funds for possible future compensation are extracted? It will not. On this account, what creates liability is risk-imposition. That compensation extracted in the name of this liability never becomes appropriate does not thereby undermine the liability itself. In any case, it is hard to see how it could be claimed with certainty that a risk created in the past at t-n will never materialise, given that t+n remains ahead. The risk may have diminished, but the existence of the future means that it will not have disappeared, and thus it is possible that victims will come to exist in the future who have a claim for compensation in virtue of the harm they suffer as a result of the ripened risk. The only point at which it can be claimed with certainty that a risk will never materialise is when no more human beings exist for whom it could materialise.

I am not suggesting that all savings made to compensate future generations be held until *Homo sapiens* goes extinct, even when the probability of the risks creating the liability that generated the compensation payments have diminished to miniscule value, but have not actually materialised. Of course, not all climate change risks are in this category: some are strongly uncertain, and will remain so. But with respect to risks to which probabilities can be assigned, at some point it might be prudent for a future generation to set aside risks imposed on them by previous generations by waiving their right to compensation, should the risk ripen into a harm. Setting the threshold of probability below which a generation should decide to waive the right to compensation attaching to that risk could be undertaken through democratic deliberation in consultation with experts using the same procedures for threshold setting with respect to taking precautionary steps against risks, as laid out in the last chapter. Any generation that has taken this decision could redistribute funds previously earmarked for compensation for harm arising from Risk R to a fund, or funds, earmarked for compensation for harm arising from Risk R1, or R1 … n. Again, decisions about the redistribution of funds will (and ought to) depend on the values and priorities (constrained, of course, by principles of justice) of the generation in question, and should be made as a result of democratic deliberation. And it could be that fund R1 (or one of R1 … n) responds to that generation's *own* liability to future generations as a result of risks it imposes on them; that is, a generation could decide to redirect funds from R1 into R2, which responds to that generation's liability to future generations in virtue of the risks it imposes on them. But whether or not this is the case, this use of R does not necessarily do any injustice to the generation who started the fund: their liability was complete the moment they imposed risks on future generations through their actions, regardless of what the future actually holds.

The corrective justice approach to intergenerational justice sketched so far raises many questions. In the section 'Impermissibility' I shall discuss how the impermissibility of climate change risk-imposition in particular is derived. In the section 'Liability' I consider the role of strict, and of fault, liability for the imposition of risks of climate change harm in particular in the intergenerational context, and the possibility of justifications and excuses for acts that prima facie underwrite the imputation of liability to their agents in virtue of their impermissibility. The sections 'Distributive injustice', 'Lack of freedom' and 'Cumulative emissions' consider three sets of circumstances in which such justifications and excuses might be thought to exist. And the 'Conclusion' surveys the ways in which an Intergenerational Climate Change Compensation Fund might function so as to realise intergenerational corrective justice with respect to climate change.

Impermissibility

Why does risk-imposition provide grounds for liability? It could be claimed that the imposition of risk, where the victim is aware of it, is in itself a harm for which corrective justice demands repair. Knowing that my chances of suffering a harm have been increased by another's acts or omissions putting me at risk places me in an environment of insecurity which depletes my well-being, and makes my claim for compensation against whoever did it legitimate. The advantage of this approach – at least for most theorists of tort law – is that it obviates the need to adjust the conception of liability in terms of cause-in-fact, i.e. it retains the traditional *ex post* approach: those who impose risks on others cause them harm, under certain conditions.

Ignoring the independently attractive reasons for adopting an *ex ante* approach to liability outlined in Rawlsian terms towards the end of the preceding section, this approach suffers from a flaw hidden in the qualifier 'under certain conditions': risk-imposition counts as a harm only if the person at risk is aware of the risk.[38] But why should persons at risk but not aware of it be denied a claim for compensation on the basis of their limited knowledge? Surely many such people suffer brute bad luck not experienced by those aware of the risks they stand under insofar as the latter can at least take precautions against the harms they are at risk of? This is true of the unborn: they cannot be aware of the risks presently existing people create for them, and so can have no claim against them for compensation. Some of the unborn will be born before those who have put them at risk die, and some of these will come to awareness of the risks under which they stand, and thus have claims of corrective justice against their ageing forbears, on this view. But this will be true of only a small subset of the unborn: those not privileged in terms of information, and (a larger class) those in the more distant future, are denied repair.

It might be claimed, as a last resort, that the distinction in desert between those aware and unaware of risks imposed on them is morally justified because the former group bear psychic costs – sleepless nights, constant worry – that the latter are free from. Taking this line, however, corrective justice is in danger of becoming the preserve of Woody Allens, constantly seeking out hidden horrors to feed neuroses grounding their claims for compensation.[39]

What should be said instead is that (some) risk-imposition is a wrong, whether or not it harms: a person put at risk of harm by another is (sometimes) wronged by her, whether or not she is aware of the risk. This is in broad terms what I am claiming with respect to the imposition of risks of climate change harm: when A puts B at risk of climate change harm, all else being equal, A is liable to provide funds to compensate B (if necessary) in virtue of her *ex ante* responsibility for the risk-imposing act, and B may claim compensation if and when the risk ripens into a harm.

Can we say, in general, what makes imposition of a risk impermissible? Note that there are at least two ways to interpret this question. First, we might want to know, in a substantive sense, *which* risks it is permissible to impose and which risks it is not. Life is full of risks, and it is impossible to avoid imposing them on others; even by sitting quietly at home one may be putting others at risk, especially if omissions as well as acts are vehicles of risk-imposition. At one end of the spectrum, we may think it is obviously impermissible to put the life of another at risk without their consent; and at the other, we may think it is obviously permissible to risk breaking someone's heart by starting a relationship with them. But what are the points on the spectrum in between? And what principles determine the point on the spectrum at which risks transform from permissible to impermissible?

I shall not attempt to give a comprehensive theory of types of impermissible risk here; my aims are less ambitious, but more significant. I have argued – using Rawlsian resources – that any current generation is not permitted to put future generations at risk of serious climate change harm, because the worst-case scenario into which these risks could ripen would deprive future generations of the possibility of justice by imposing on them unbearable strains of commitment. It is the imposition of this substantive risk upon which I focus here. Nevertheless – and with reference to the second sense of the question about impermissibility mentioned above – it remains necessary to ask: what properties of an act imposing risks of climate change harm makes this act impermissible? Properties of the act itself, independent of its possible consequences? Properties it has in virtue of its possible consequences? Or, properties it has in virtue of the actual consequences it will have? At an abstract level, what is it about, for example, emitting greenhouse gases (and thereby imposing risks on others) that makes these acts impermissible?

Judith Jarvis Thomson outlines an attractive first approach to these questions. She begins with the question: is it impermissible to cause the harm H (death, say) at which a person P is put at risk by an act A? If it is impermissible to cause H, then we might be tempted simply to claim that if A will in fact cause H, then A is impermissible (and if A will not, in fact, cause H, then it is permissible). Thomson describes this approach as an instance of an 'inheritance principle' at work in the moral evaluation of risk: 'If A ought not to cause B's death, then if it is the case that if A verb-phrases, he will thereby cause B's death, then A ought not verb-phrase … If you may not cause a man's death, then surely you may not do that which you would cause his death *by* doing.'[40]Applied to risks of climate change harm: if emitting greenhouse gases will in fact cause serious climate change harm, then emitting greenhouse gases is for that reason impermissible.

However, Thomson notes that this approach is defective on two fronts. First, it does not capture the seeming impermissibility of imposing some risks, even when no harmful outcome is actually caused by the act imposing the risk. Thomson's example is of a person playing Russian Roulette on another (sleeping, and so entirely unaware) person: as it turns out, no harm comes to the sleeper, but we would still want to say that the Roulette player did something impermissible in imposing a one-in-six chance of death on the sleeper. Second, and with greater relevance to the case of climate change, the approach makes it impossible to judge in advance the (im)permissibility of acts we would, in the normal course of our decision making, judge to be (im)permissible. For example, there are various types of act which we (rightly) judge to have a very small probability of causing harm: cooking for friends, using the light switch, digging the garden. However, in some rare sets of circumstances tokens of these types will cause serious harm to others: lethal poisoning, death by domestic fire caused by a short circuit, the contraction of fatal cancer as a result of exposure to toxic gases released by the spade in the soil. A story can be concocted about the effects of any token act, however seemingly risk-free the type, whereby it stands as the cause of a serious harm. Even by sitting quietly at home I could be putting others at risk of serious harm, especially if omissions as well as acts can generate moral responsibility: had I chosen to go out that day instead I might have saved a child from death by pulling her out of the path of an oncoming truck.[41] In any set of circumstances there may be some token acts of the type that will, in fact, have a seriously harmful effect, and it is not permissible to perform these acts, according to the approach informed by the inheritance principle that is under consideration.

But, then, how are we to decide in advance of performing any act whether it is permissible, especially when what we think about the question – that is, our judgements or knowledge of the risks involved – is irrelevant to its answer. Part of the point of thinking about permissibility is that it aids our decision making about whether or not to perform an act, in advance of taking steps to that end. If permissibility can only be established after performance of the act – once all its consequences have played out – then it is redundant in practical reasoning. As Thomson puts it, '[w]e do not think that the permissibility of acting under uncertainty is to be settled only later, when uncertainty has yielded to certainty'.[42] With respect to climate change, this approach might not provide a solid foundation for making greenhouse gas emissions impermissible because of the levels of uncertainty in climate change science, as well as the huge difficulties involved in establishing that any particular harmful event (for example, a tropical storm) is (a) an effect of climate change, which is (b) attributable to past greenhouse gas emissions. I shall have more to say on these difficulties later on.

Thomson identifies two ways out of this dilemma, without endorsing either. First, that the inheritance principle be abandoned, meaning that even risky acts certain to have seriously harmful effects could be permissible, *ceteris paribus*. Taking this approach, all the work of determining permissibility is done by the *ceteris paribus* clause. At the very least, and prima facie, two features of an act to which this clause relates will be that its agent does not intend the seriously harmful effect, and is ignorant of it. However, we can imagine cases in which the prima facie nature of this claim is undermined: perhaps seriously harmful acts known to be and intended as such by their agents are permissible in cases of self-defence.[43]

The second way out of the dilemma is to think of impermissibility as attaching directly to the imposition of risk itself, and not to any bad outcome it may or may not have. This is the more radical strategy. On this approach, it is (for example) impermissible to impose a high risk of death on someone, but not impermissible actually to cause their death. The most we can say is that causing another's death is impermissible if and only if the action that causes the death imposes a high risk of death. And not all actions causing actual deaths do impose a high risk of death (remember cooking dinner for friends, using the light switch, and digging the garden). On this view, an inheritance principle operates, but it works in the opposite direction, so to speak, to the one operative in the dilemma-creating approach. Actions are impermissible if and only if they impose a high risk of serious harm (*ceteris paribus*), and not in virtue of standing as the cause of an actual serious harm.

Thomson rejects this approach on the grounds that it is too counter-intuitive to adopt a view with the consequence that 'it just never is true to say of anyone, whatever the circumstances, that he ought not cause a person's death'.[44] However, it is unclear why she includes the qualifier 'whatever the circumstances', and it is this that clashes with our intuitions. We do not want to *have* to say that, whatever the circumstances, a person ought not to cause another's death: we want to make room in our moral assessment of acts for cases of permissible killing (although of course what these cases are is a hard question). But the approach just outlined does just that: it does not imply that causing death can never be said to be impermissible. It just implies that those cases that are impermissible inherit their impermissibility from the fact that the act that in fact caused them was high risk with respect to the harmful outcome.

Thomson's objection to this approach is that we cannot say of these cases *directly* that they are impermissible: the impermissibility of a harmful outcome is always a function of the impermissibility of the risky act that causes it. This mirrors an objection to the approach in which the inheritance principle worked in the opposite direction, and explained the impermissibility of risk-imposition in terms of the impermissibility of the harm giving content to the risk (the Russian Roulette case): highly risky acts, however knowingly performed with whatever horrible possible consequences, can never be said to be impermissible unless they mature into an actual harm. In both sets of cases, the operation of an inheritance principle delivers counter-intuitive judgements about the permissibility of key types of risky, and actually harmful, acts.

Perhaps, then, abandoning the inheritance principles that generate these counter-intuitive judgements is, after all, the best option. This would mean finding accounts of the impermissibility of harmful outcomes, and risky acts, which are independent of one another. With respect to my interests here, it would mean finding an explanation for the impermissibility of imposing high risks of climate change harm on future people through greenhouse gas-emitting activities that does not depend on the claim that these harms are certain to come to pass. I offered just such an account in the last chapter, which made appeal to the duties of justice any generation has to future generations to deliver the circumstances of distributive justice to them, and not to risk imposing unbearable strains of commitment on them. A generation has these duties at any point in time in the face of uncertainty about the effects of their present acts on future generations, and the performance of these duties remains obligatory even if it will in fact come to be the case that future generations are not harmed by climate change, and would not have been harmed even if the duties had not been performed. Here, the explanation of the impermissibility of not performing these duties depends on reflections about what any generation could itself accept by way of principles to govern acts by generations preceding it (this is the function of the original position in thinking about intergenerational justice), and not on the certainty that impermissible harmful outcomes will come to pass for future generations if these principles are not adopted.[45] Thus, there is no inheritance principle at work in this account of impermissible risk.

Moving on, if risk creation is a wrong and not a harm then the following objection arises: principles fit to address the imposition of risk are not principles of corrective justice, but rather higher-level principles of justice, accompanied by principles of retributive justice that ensure appropriate penalties for the wrongdoing (which could, of course, take the form of a fine to be paid to an intergenerational fund). If A's Ø-ing wrongs B, then (*ceteris paribus*) we ought to require that A not Ø, and institute prohibitive laws, with penalties for violation attached for anyone who Øs, rather than – as the corrective justice approach has it – requiring just that A make a contribution to a fund to compensate those harmed by Ø-ing. The corrective justice approach, goes the objection, fails to register the wrongness of the acts which it treats as grounds for liability, but adequately registering this wrongness in principles of justice would make corrective justice redundant, because there would be nothing to correct for. Applied to climate change, the claim is that corrective justice fails to provide the ideal solution to the problems of intergenerational justice created by greenhouse gas emissions; rather, if emissions are unjust to future generations, they ought to be prohibited.

There are two lines of response. First, if Ø is a wrong, and is reasonably avoidable, then (*ceteris paribus*) it ought to be prohibited, and in the world of ideal theory where compliance with laws is strict, that would be, prima facie, the end of the matter, with no further role for corrective justice. But climate change is a nonideal problem par excellence, and the world contains few institutional mechanisms to enact and enforce international and intergenerational prohibitive legislation fit effectively to regulate the activities that cause it. In the absence of this, and the likely real-world non- and partial compliance with requirements of justice as they would be expressed in prohibitive legislation related to greenhouse gas emissions, the corrective justice approach at least ensures that victims can get some compensation, even if it does not visit upon the wrongdoers the punishment they deserve. That the world is such that wrongdoers can escape punishment should not be held as a reason for doing nothing to help the victims that their wrongful behaviour creates as the risks it imposes mature into harms.[46] And if victims will come into existence long after the wrongdoers cease to be, then a good way to ensure (some) justice for the victims is to extract funds for reparation from the wrongdoers at the point of their wrongdoing, while at the same time insisting that, ideally, they should stop doing wrong.

The second line of response is this. Even in an ideal world of strict compliance with just laws, people can put one another at risk of serious harm through their actions. Not all wrongdoers do wrong intentionally, knowingly, deliberately, etc., and even those who strive to meet appropriate standards of care in what they do may sometimes fail, and not always through their own unintended negligence. With respect to climate change, the science is incomplete, and lags behind the phenomenon, making levels of uncertainty about many aspects of it high; even with the best intentions, people will inevitably make mistakes about what is best to do, and how it is best done. Unless the agents of ideal theory are omnisciently prescient with respect to the harms their actions could cause, and are omnipotently able to ensure that mistakes and 'moments of carelessness' will never happen, we should accept that even in an ideal world wrongful acts imposing risks of serious harms will be performed. As was emphasised in Chapter 1, it is not the purpose of ideal theory to idealise away all human failings, organisational limitations, or conflict arising from social interaction: the people of ideal theory are not angels or gods. Agents who harm one another as a result of their 'moments of carelessness' are not necessarily criminally liable; but if, as per the objection just considered, prohibitive principles exhaust the normative resources available to specify how people in ideal circumstances ought to live together, then there will be nothing left to say about who should bear the costs when a risk of serious harm imposed by one person on another matures into such a harm for the victim; and these costs will not go away. Even in ideal circumstances of strict compliance, there is a place for corrective justice. With respect to climate change and intergenerational relations, *ex ante* responsibility, and the liability for risk-imposition that it generates, stands as a form of insurance against the possibility that – in a context of strong uncertainty about how what we do affects the climate – actions are being performed that will cause harm to future generations, despite the best efforts of agents who strictly comply with principles of justice.

Liability

Returning to the nonideal world, the imposition of risks of serious climate change harms on future generations by the current generation is widespread and, given the state of climate change science, not an unforeseeable consequence of emissions, or an unavoidable mistake. With a theoretically well-grounded account of intergenerational liability now in place, can we say that the current generation's risk-creating greenhouse gas emissions are sufficient to establish their liability according to intergenerational corrective justice?[47]

At this point let me introduce a distinction from tort law between strict and fault liability. As I have said, my aim is not to give an account of corrective justice which makes sense of tort law, but rather to articulate a conception of corrective justice which provides just solutions to the problems of intergenerational justice created by climate change. However, the interpretation of tort law in terms of corrective justice is prominent and influential, and discussion of the two categories of liability operative in tort law – interpreted in terms of corrective justice – is necessary in order to answer the question about liability posed at the end of the preceding paragraph. It will also prove illuminating.

Following Jules Coleman, strict and fault liability can be distinguished as follows. An agent is liable for a risk she imposes on grounds of fault when the risk-imposition is wrongful in virtue of how it breaches a duty of justice the agent has towards those on whom she imposes the risk, where the duty requires a standard of care that the agent fails to meet in virtue of doing something she ought not to have done, or not doing something she ought to have done. In contrast, an agent is strictly liable for a risk she imposes on others when the risk-imposition is wrongful just in virtue of how it breaches a duty of justice the agent has to those on whom she imposes the risk, where this duty need not include a specification of the standard of care an agent must satisfy in order to perform the duty.[48] A driver who takes her eyes off the road for a moment and hits a motorcyclist is liable on grounds of fault: the duty that she breaches is the duty not to drive carelessly. A lion tamer whose lion escapes from her circus and kills a child is strictly liable, regardless of the quality of her cages and procedures used for monitoring the animals: the duty that she breaches is the duty to prevent her lions from killing anyone. It is consistent with her strict liability, in this case, that her cages are weak and her procedures flawed; and she may also be liable on grounds of fault for these reasons. However, these facts need not be established in order for strict liability to be ascribed to her. As Coleman puts it, 'Fault, especially negligence, liability requires fault in the doing, but not in the doer. Strict liability requires neither'.[49,50]

Let me comment on the strict/fault liability distinction so made before applying it to the intergenerational context. As a first stab, Coleman puts it as follows,

> Reasonableness is a defense against fault liability. To show that what one did was in the circumstances reasonable is to show that what one did was reasonable or permissible. Thus, ascriptions of fault liability are defeasible by justifications, if not by excuses. We can then define strict liability in torts as liability that is defeasible *by neither excuses nor justifications*.[51]

A justification exempts a token act from the impermissibility that attaches to its type; an excuse highlights features of the agent in virtue of which she ought not to be held liable despite the impermissibility of the token act she performs. Prima facie, excuses have an important role to play in the criminal law but not in the law of torts. This is because both fault and strict liability attach to agents as a result of what they do rather than of their character as doers of these things. However, this shows that the prima facie irrelevance of excuses to tort liability really is prima facie: some excuses aim to show that an agent did not, in fact, perform the impermissible act in question, and these excuses need bear no relation to the character of the agent as a doer. Thus (as Coleman goes on to affirm) both excuses (of the right kind) and justifications can function to relieve an agent of fault liability when her action causes harm to, or (in our case) imposes risks of harm on, another either by showing that the action was not wrongful, despite the impermissibility of the type of which it is a token (as in the case of justifications), or by showing that the injurer was not in fact the agent of the impermissible act, or ought not to be held liable in virtue of some other features she possesses (as in the case of excuses). I discuss some cases relevant to climate change in the next three sections.

For the moment, note that the indefeasibility of strict liability by excuses or justifications means it is consistent with the ascription of it to an agent that her actions were exemplary: it may be that the lion tamer's cages are the best that money can buy, and that her procedures for monitoring her lions are impeccable. This feature of strict liability could suggest that it has no role to play in any system of just rules governing social interaction, because agents can be held strictly liable for harms they cause (or risks they impose) when they are blameless, and their actions are faultless, and this is prima facie unfair.

Coleman's account of how fault liability also has at its heart a strict allocation of costs when one agent harms another, or puts her at risk, dispels the appearance of unfairness in strict liability. He starts with the thought that when A harms B, costs are created that will not go away: someone must bear them. Strict liability rules allocate all these costs to A, all else being equal. However, fault liability rules also strictly allocate these costs when no fault can be established; in these cases, it is B – the victim – who must bear them. As he puts it,

> The rule of fault liability … has a rule of victim liability contained within it, and it is a rule of strict liability. If we wanted to capture completely the rule of fault liability, it would read something like this:
>
> […] The victim is strictly liable for her losses unless she can establish the fault of her injurer in having occasioned them.[52]

Furthermore, the strict liability in strict liability is more defensible than the strict liability in fault liability. The agent held strictly liable under strict liability is the cause of another's losses, whereas the agent held strictly liable under fault liability when no fault is established – i.e. the victim – is not the cause of her losses.

How might an agent defend herself against strict liability? Apart from showing that the relevant duty of care was not violated, and/or that causation was absent, strict liability can be rebutted if the victim was at fault as in, for example, contributory negligence. Thus, there are actually two versions of strict liability: strict liability with the defence of victim fault, and strict liability without the defence of victim fault. In a very early paper Coleman refers to the latter as 'absolute injurer liability',[53] and I shall follow him in this. It is worth noting now that if strict liability applies in the intergenerational context to any generation imposing impermissible risks of climate change harm on its successors it cannot be with the defence of victim fault, as victims either were children when members of a previous generation imposed risks on them, or were not born. Thus, in thinking about intergenerational liability we must choose between fault liability and absolute injurer liability.

We can now state more precisely the question with which we opened this section. Should fault or absolute injurer liability attach to any current generation putting future generations at risk of serious climate change harms? Or should different types of liability attach to different groups within any current generation? And if so, why?

I have argued so far as if at least some of the current generation have absolute injurer liability for the risks of climate change harm they impose. However, it is now time to give this assumption closer attention by addressing the following question: have some, or all, of the current generation breached a duty to future generations not to faultily place them at risk of serious climate change harms? If the answer to this question is 'yes' then some or all of the current generation are liable on grounds of fault. Assuming for ease of exposition that the imposition of climate change risks is effected just by emitting greenhouse gases, answering this question depends on considering whether it is possible to emit greenhouse gases in a non-faulty way. If this is not possible, then liability for all emissions is absolute. What is really at issue here is whether there are justifications and/or excuses for emitting greenhouse gases which render them non-faulty and so not in breach of any duty of intergenerational justice or, if in breach of such a duty, excusably so.

Justifications and excuses function in law in general to exempt an agent from liability for the performance of a prima facie impermissible act. Justifications draw attention to properties of the token act in question, and features of the circumstances in which it is performed, which require or permit that it be exempted from the prohibition applying to other tokens of that impermissible type. Excuses draw attention to properties of the agent of an impermissible act, or features of her circumstances, which defeat the ascription of liability to her either by relieving her of culpability ('blame defeating excuses') or by showing that she was not the author of the act ('agency defeating excuses').[54]

When an act is excusable, the type and the token in question are impermissible, but the agent is not liable because facts about her circumstances (including her personal situation) either sever the connection between the wrongness of the act and the culpability of its agent, or show that her agency was not genuine. In contrast, when an act is justified, the type is impermissible, but the token is not. Because liability rules in tort law attach to agents just in virtue of their authorship of wrongful acts, and not in virtue of their culpability qua agents, blame-defeating excuses have no role to play in the determination of tort liability. This is one reason why it is plausible to interpret tort law as aiming at corrective justice: under the correlativity requirement for corrective justice, an agent must be responsible for the injury her action causes to another, but responsibility is not specified in terms of culpability.[55] Of course, sometimes people are culpable for the injuries they do to one another; indeed, sometimes these injuries are inflicted intentionally, deliberately, maliciously, or with forethought. The relevant principles in these cases are principles of retributive justice, at which criminal law aims. Here, there is a role for excuses: for example, criminal liability can be defeasible by showing a defendant is not of sound mind.[56] However, because correlativity requires that the injurer be responsible for the harm suffered by the victim, agency-defeating excuses do have a role to play in the determination of tort liability, interpreted in terms of corrective justice. If a person is not the agent of a wrongful act – if, for example, she acts as a result of having been brainwashed or drugged – then she has an agency-defeating excuse in virtue of which she should not be held liable for any harm done to a victim by the act.

My interest here is in the possibility and nature of justifications and agency-defeating excuses for the imposition of climate change risks on future generations as they function in corrective justice. If such justifications and/or excuses exist, then this is interesting for at least two reasons. First, and insofar as the categories of tort law aim at the ideals of corrective justice, it would show that the appropriate form of liability for the imposition of these risks is fault based, because if the appropriate form of liability were strict there would be no convincing cases of justifications or excuses for such risk-imposition (remembering that strict liability is defeasible by neither excuses nor justifications, and fault liability is defeasible only by justifications and agency-defeating excuses). Second, it would show, substantively, which groups ought to be targeted by any institutional scheme for the realisation of intergenerational corrective justice with respect to climate change. I discuss the second interesting implication in the final section. In the next three sections I consider cases in which it is plausible to claim justifications or excuses exist, and which relate to the first interesting implication. These cases are made by reference to distributive injustice, lack of freedom, and cumulative emissions.

However, before moving on it is worth mentioning a third way to think about liability for risks of climate change harm in the intergenerational context, in order to set it aside. With respect to a large class of risky activities, the benefits associated with them make it desirable that we should be able to engage in these activities without liability for contributions to a fund to compensate possible victims, even taking into consideration the risks these activities impose on others. Many everyday activities carrying a small risk of harm have this character: driving a car (within the speed limit, and with due care and attention) is a good example. As McCarthy suggests (plausibly) with respect to such activities, we can be thought of as thereby imposing 'reciprocal risks' on one another, which has the following consequence for liability: 'since the magnitude of the risks [we] impose on each other are the same, we could regard these duties [to provide compensation – CM] as cancelling out. In short: no liability for reciprocal risks.'[57] Are greenhouse gas emitting activities – or a significant subset of them – of this type? Can we be thought of as imposing climate change risks of similar magnitude on one another, thereby cancelling out any liability we otherwise have to one another?

There are at least two reasons to reject this approach with respect to climate change risks. First, much would have to be done to show that the greenhouse gas-emitting activities through which parties impose reciprocal risk upon one another carry sufficient benefits for all so as to justify appealing to reciprocal risk to cancel out liability. There are many who would question this, for example, the inhabitants of the Maldives, Kiribati, or other small island states. In addition, those who advocate a 'no regrets' strategy of justification for cutting emissions would question the overall benefits of such activity, arguing instead that even if the risks created by emissions are reciprocally imposed there are still great benefits for all to be gained from cutting emissions, giving independent reasons for doing this, whatever is true about reciprocal risk-imposition in this area. Second – and most importantly, for my purposes – in order for this claim about reciprocal risk and liability to hold, there has to be a context for reciprocity – and the context of intergenerational relations does not provide this. We can impose risks on them, but they (most of them) cannot impose risks on us. In the intergenerational context, using reciprocal risk-imposition to cancel out liability is a non-starter.

Distributive injustice

The case for justifications or excuses related to distributive injustices can be stated as follows:

> The least-advantaged in any unjust distribution of goods are already suffering from injustice. To hold them responsible under corrective justice – and liable under any institutional system for its realisation – for their emissions would compound this injustice by making them even worse off, and warrants treating them as if they cannot perform faulty acts of emitting.

This case describes circumstances in which justifications ought to obtain: the risk-imposing acts of those suffering this injustice *ipso facto* ought to be exempted from prohibitions applying to other tokens of the impermissible type. The claim is that distributive injustice provides a justification for relieving those suffering from it from liability. It is true that according to corrective justice, a person's liability and/or entitlement to compensation relates to how she has acted on/been affected by the actions of others, and not to the position she occupies in any distribution of goods, just or unjust. Crudely, we do not determine whether a person deserves compensation for harm done to her by considering whether the liable agent can afford to pay compensation, or whether the victim already has her fair share according to principles of distributive justice: under corrective justice, Bill Gates can be just as deserving of recovery as Mother Teresa.[58] This feature of corrective justice suggests that it can sometimes require transfers that disrupt distributions of goods conforming to principles of distributive justice; that is, that it can demand transfers that transform just distributions into unjust distributions or, as per the objection, that it can demand transfers that transform unjust distributions into even more unjust distributions. If this is true, then the perspectives of corrective and distributive justice could not – as I claimed in the Introduction – be complementary parts of any incompletely theorised agreement on low-level principles to govern intergenerational relations: in defending these principles we would have to choose between the perspectives of corrective and distributive justice because they issue in contradictory injunctions and so are not consistent. Thus, given that I aim to present the perspective of corrective justice in a way fit to form part of such an incompletely theorised agreement, it is important to clarify that corrective justice need not be conceived of as requiring distributively unjust transfers.

The 'distributive injustice' case can be read in at least two ways: (1) that distributive injustice is worsened when and because the worst-off are made even more worse off; (2) that distributive injustice is worsened when liability is ascribed to the worst-off, because an ascription of liability is an accusation of having acted unjustly.

Let us start with (1). Although Bill Gates can be deserving of recovery, it is consistent with corrective justice that the funds to satisfy his claims for recovery issue from a source other than Mother Teresa altogether, even when Mother Teresa is liable for the harms he suffers. This is the case for the following reasons: (i) A can be liable to B, and (ii) B can be deserving of compensation in virtue of A's liability to her, even when A and B do not stand in a relationship of distributive justice. But note that (i) and (ii) do not entail (iii) A must provide compensation to B. When A and B are not in a relationship of distributive justice it may be legitimate for us to exempt A from the requirement to compensate B, while still holding that A is liable for the harm done to B, and that B deserves compensation for this harm. When A is very poor, and B is very rich, and there is no distributive justice between them, we may decide to fund B's compensation from the public purse on the grounds that requiring A to meet the obligation to compensate B generated by her liability to B would compound a distributive injustice from which A already suffers. It is possible to hold that (i) and (ii) can be true of parties whatever their circumstances, and yet insist that (iii) holds only when A and B stand in relations of distributive justice. This matters for intergenerational justice because it permits us to exempt liable but very poor people in any current generation imposing risks of climate change harm on their successors from the requirement to provide funds for recovery. These institutional arrangements are perfectly compatible with corrective justice.[59] As I have said, I help myself to the categories and concepts used in tort law insofar as they are illuminating and consistent with the vision of corrective justice I am articulating. The requirement in tort that the injurer provide funds to compensate the victim is only one way of fulfilling what Coleman calls the 'rectification' condition for corrective justice (see p. 74). If transfers of this sort would entrench, or worsen, distributive injustices, and if we are committed to a vision of justice in which distributive justice plays a key role (which is a fundamental Rawlsian commitment), then we should endorse alternative methods for securing funds fit to compensate victims. More will be said on this topic in the Conclusion.

One objection to this suggestion relates to the functions of compensation. It matters to victims that the party liable for the harm they suffer is the same party that provides compensation: compensation matters to victims not just because of its use-value to them, but also because of its symbolism when extracted from a liable party. Compensation is a tangible symbol of the liability of the party from whom it is extracted, and its award to a victim is a public recognition of their suffering. My suggestion that compensation could sometimes be funded from the public purse, even when a liable party has been identified would, the objection goes, undermine the symbolic value of compensation.

In response, it is true that the delivery of compensation can have symbolic value for victims. However, other acts performed by a liable party could also have such value, and perhaps more potently. Perhaps the requirement that liable parties from whom compensation is not extracted should make a public apology to their victims would satisfy the legitimate desire of victims for an acknowledgement of their righteousness according to corrective justice. Further, perhaps this requirement ought to hold universally with respect to liable parties, whether or not they fund compensation payments to victims, i.e. all liable parties could stand under a requirement to make public apology, and a subset of these – determined by considerations of distributive justice – would also be required to make compensation payments. However, making public apology a requirement of corrective justice risks conflating one of its functions with a function of retributive justice, that is, that of punishment of a liable agent. If public apology is to be included as a requirement of corrective justice then efforts must be made to avoid public humiliation of the liable agent; that is, to avoid the ceremony of apology becoming a virtual stocks.

Turning now to the second way of reading the 'distributive injustice' case, the claim is that distributive injustice is worsened when liability is ascribed to the worst-off, because an ascription of liability is an accusation of having acted unjustly. The objection thus stated is a non sequitur: an ascription of liability is indeed an accusation of having acted unjustly, but it is not clear why this compounds an existing distributive injustice unless, perhaps, the accusation is unfounded. It should go without saying that unfounded ascriptions of liability are unfair to those who suffer them, and so should be avoided. But it does not follow that those who are distributively disadvantaged, even unjustly so, are thereby treated more unfairly by an unfounded ascription of liability than those who are (justly or unjustly) more advantaged. And nor does it follow that the unfairness suffered in this case compounds the *distributive* injustice; instead, the false accusation stands as an independent injustice done to the victim. For these reasons, I dismiss the second interpretation of the 'distributive injustice' case. And because the first interpretation does not establish that those suffering distributive injustice are not, for that reason, in a position to faultily emit greenhouse gases, the 'distributive injustice' case as a whole does not give us reason to conceive of liability for the imposition of risks of climate change harm as fault based.

Lack of freedom

I suspect that when the 'distributive injustice' case is offered as grounds for a justification of greenhouse gas emissions what is really meant is the 'lack of freedom' case. The case for justifications or excuses defeating liability for emissions can be stated as follows:

> Different levels of advantage should affect our judgement as to whether a putatively liable person could have acted differently; broadly speaking, people with fewer resources have fewer options for acting differently than those with more resources. A person who has no option but to act as she does does not act in a faulty way.

Prima facie, this case describes circumstances in which agency-defeating excuses ought to obtain: people in poverty are not the authentic agents of their risk-imposing acts. This case raises the question of whether different levels of advantage should affect our judgement as to whether a putatively liable person could have, or could reasonably have been expected to have, acted differently. Surely the ageing rock star who chooses to fly his favourite hat first class from London to Italy so he can wear it on stage,[60] and a single parent on a benefits budget buying carbon-intensive cheap food to feed her kids, should not be tarred with the same brush of liability, even if (as is clearly not the case) there is no distributive injustice between them?

It is true that for many people in poverty the costs to them of reducing their emissions would be so great that to treat them as being in a position to act otherwise – as Bono indisputably was – would be grotesque. For a young single parent in Michigan dependent on the Temporary Assistance for Needy Families (TANF) programme in the US, installing cavity-wall insulation is not an option. For a day labourer in Kohlua, central India, living without electricity and earning $2 per day, there is no choice but to use dung cakes and wood on cooking stoves that emit soot – the so-called 'black carbon' that could be responsible for 18 per cent of global warming – into the Asian 'brown cloud'.[61] Or, perhaps more problematically, does a teacher in an Indonesian village who supplements his income with a smallholding producing palm oil have the option to do otherwise, given that this would mean sinking his family below the poverty line?[62]

However, there is a problem here. Agency-defeating excuses aim to show that the injurer did not, in fact, have agency in performing the wrongful act inflicting harm on the victim; agency-defeating excuses show that the putative agent did not act at all. The examples given by Coleman are '[m]oving one's body under a hypnotic trance' and shooting someone as a result of having had a device placed in one's brain by someone else, and involuntary twitches.[63] With these examples as illustrations he states the rationale behind agency-defeating excuses as follows:

> [E]ven if an individual can be held liable for what she has done, whether or not her doing it is culpable in any way, she cannot be held liable in justice for things that, although they involve her, happen to her, not by her.[64]

The circumstances of the 'lack of freedom' cases – as illustrated by the examples of the American single parent, the Indian day labourer, and the Indonesian teacher – are not those in which agency-defeating excuses so conceived obtain. It is not the case that poverty robs those suffering from it of the capacity to act per se, however this capacity is conceived. Poverty certainly severely limits the options available to those in it, and so restricts the scope of their agency, but agency remains nonetheless. Thus the 'lack of freedom' case does not establish the existence of agency-defeating excuses for those in poverty.

Although poverty does not deliver agency-defeating excuses, reflecting further on the emissions of people in poverty will uncover a solid basis for differences in the ascriptions of liability to poor and rich emitters. Consider as a starting point this reasonable response to the Bono example: the difference between Bono and the Indian day labourer is not that Bono has the capacity for agency and the day labourer does not. Rather, it is that the costs to Bono of acting otherwise make it reasonable to expect him to act otherwise, and the same cannot be said of the day labourer. Bono's feeling of insecurity in not having his favourite hat on his head when singing with Pavarotti is not of comparable moral significance to the day labourer's children going without food, or without heat in the winter months.

To put flesh on the bones of this intuitive response, consider Henry Shue's distinction between 'luxury' emissions and 'subsistence' emissions: 'some sources [of greenhouse gases] are essential and even urgent for the fulfilment of vital needs and other sources are inessential or even frivolous'.[65] He uses this distinction in the following argument. In order to avoid exceeding the ceiling on the planet's atmospheric capacity to absorb greenhouse gases, reductions in total emissions are necessary. Current energy regimes mean it is impossible for anyone to survive without emitting some greenhouse gases. To require the world's poorest people to cut their subsistence emissions (or even to require that they not increase their emissions) in order to achieve these reductions 'would be unfair to the point of being outrageous' because the reductions could instead be made by richer people cutting their luxury emissions. As he puts it,

> These are not two different means to the very same end. The ends of the two different measures could be the same in the amount of greenhouse gas emissions they eliminate, but the ends are otherwise as different as reducing vital supplies of food and making luxuries a tad more costly.[66]

Why would it be 'unfair to the point of being outrageous' to require the poor to cut their subsistence emissions in order that the rich can retain (or expand) their luxury emissions? Shue argues that the limited capacity of the atmosphere to absorb greenhouse gases, combined with the fact that '[f]or practically everyone at present, and for the immediate future, survival requires the use of greenhouse gas emissions' absorptive capacity',[67] means that the absorptive capacity of the atmosphere should be treated as a scarce resource essential for survival in the same way as food or water. On any reasonable conception of distributive justice, when a resource is scarce and essential for survival it ought to be distributed so that as many people as possible can survive. In particular, amounts of this resource ought not to be allocated to anyone in order that she can use it for purposes other than survival when there is not sufficient, or only sufficient, in total for each to have the amount necessary for survival: in a zero-sum game of distributing scarce survival resources, an allocation such as this to one person will have the result that another person dies. Seeing things this way, luxury emissions by some are possible within the limited absorptive capacity of the atmosphere if and only if others are denied subsistence emissions, which means death for them.

Shue's argument helps in pinpointing what it is about the circumstances of people in poverty that makes us worry about ascribing liability to them. Because of their poverty, all their emissions are subsistence emissions: in holding them liable for these emissions it looks like we are endorsing a prohibition which, if enacted, would require them to choose death, and this cannot be acceptable. Note, though, that what makes it possible for rich people to be luxury emitters is that they too use up a portion of the atmosphere's absorptive capacity in subsistence emissions: without being alive they would not be in a position to make further emissions in their pursuit of luxuries. However, what any rich person emits in order to feed, water, and shelter herself and her family is likely to be far higher than the emissions of any poor person because of how the rich feed, water, and shelter themselves. For example, the air miles attached to imported food have a vastly larger carbon footprint than a fire made of dung cakes on which home-grown rice is cooked. Nevertheless, as Shue emphasises, our current energy regime makes some emissions in the pursuit of survival an 'avoidable necessity'. It is avoidable because emissions need not be – and for much of human history, have not been – the way in which survival is ensured; and it is a necessity because, given our global energy regime, there is no alternative but to emit in order to survive. As I have argued throughout this book, we have duties of justice to future generations to change our carbon-intensive energy regime (and pretty fast) in order that they are not deprived of the possibility of justice. However, to claim that could be done overnight is either fantasy, or entails that the majority of the current generation fall on their swords in forswearing their subsistence emissions. What this suggests is that we ought not to hold anyone – poor or rich – liable for their subsistence emissions, understood in terms of what is required in order for them to survive. That subsistence emissions are an 'avoidable necessity' relieves each member of the current generation of liability for these emissions.

However, does this fact generate a justification for subsistence emissions or (what I shall call) a 'freedom-defeating' excuse? Remember that a justification shows a token of an impermissible type to be, on reflection, permissible, whereas an excuse shows an agent to be non-liable despite the impermissibility of the token act she performs. Subsistence emissions, albeit an avoidable necessity, contribute, however minutely, to total cumulative emissions, and chip away at the atmosphere's finite absorptive capacity, leaving less and less for future generations at least 150 years hence (given the atmospheric life of CO_2) on the best-case scenario, and possibly far further into the future on worse-case scenarios. The effects of subsistence emissions on the availability of the scarce resource of the planet's atmospheric capacity to absorb any emissions makes all token acts of emission impermissible. The fact that every person on the planet has, at base level, a choice between emitting greenhouse gases to survive and death does not mean that emitting greenhouse gases must be morally permissible: emitting greenhouse gases for whatever purpose creates risks of serious climate change harm for future generations, and so is wrong. The current generation is, in this sense, facing a moral tragedy: we cannot be expected to choose our own deaths, but in avoiding death we must do wrong.

What the 'lack of freedom' case shows is that a freedom-defeating *excuse*, rather than a justification, exists for subsistence emissions. This is not an agency-defeating excuse: agency is exercised when a person chooses to refrain from subsistence emissions and, given our current energy regime, thereby choose death. But exercising agency over this narrow range of options hardly qualifies as freedom. If we were to hold every member of the current generation liable for her subsistence emissions we would be committed to requiring them to do something they are not free to do, namely, cease their subsistence emissions and choose death.

These reflections on the 'avoidable necessity' of subsistence emissions show that there is a non-faulty way to emit greenhouse gases, i.e. in the pursuit of survival, given the existing carbon-intensive energy regime. And this shows that fault rather than absolute injurer liability is appropriate for the operation of intergenerational corrective justice with respect to risks of climate change harm created for future generations by the current generation: luxury emissions are faulty emissions and those who create them are liable.

I have, quite deliberately, not quantified the acceptable level of emissions required for survival. In particular, I have not commented on whether, given how differently the current energy regime operates in various national contexts, ceilings and thresholds for subsistence emissions can be set at the same level across these contexts. There are of course some very obvious exclusions. Bono would not have died in Modena without his hat, and no one needs to eat Almas caviar to survive.[68] But is the benchmark for the limit on subsistence emissions to be set by reference to the day labourer in Kohlua? Or, rather, should the remaining available absorptive capacity of the atmosphere be divided up on a per capita basis which could mean that the day labourer is permitted to increase his subsistence emissions, but the single mother on TANF must reduce hers?[69] These questions belong to the domain of intragenerational distributive justice, where subsistence emissions form part of its currency. For my purposes, I leave open the question of how subsistence emissions are to be quantified, and what allocation would be delivered by this method. Whatever the answer to this question, my claim is that there are freedom-defeating excuses for emissions at subsistence level that relieve each member of the current generation of liability for these emissions, both in cases where their actual emissions are below this level (the day labourer, perhaps?), and in cases where these emissions are (possibly a miniscule) portion of their total emissions (Bono, definitely).

Are there further grounds on which liability for emissions could be defeated, thus strengthening the case for conceiving of such liability in terms of fault? Let me turn now to the 'cumulative emissions' case.

Cumulative emissions

The case for justifications or excuses related to cumulative emissions can be stated as follows:

> A person's emissions imposing risks on others is impermissible only because others have not curbed their emissions. Here, the faultiness of emissions is a function of more than what the agent does, and thus should not be attributed just to the agent's actions, in which case these actions cannot be faulty.

This case describes circumstances in which justifications ought to obtain: the risk-imposing acts of any individual are only impermissible because of the impermissible acts performed by others, and for that reason ought to be exempted from prohibitions applying to other tokens of this type. Consider how the case would work in a concrete example. Let us assume that the act type 'driving an SUV' is impermissible because of the risks of serious climate change harm imposed on future generations by tokens of this type. Furthermore, let us assume that there is a threshold of total emissions above which dangerous climate change becomes irreversible and risks of serious climate change harm to future generations are imposed (and also assume that we have not yet passed this threshold). For each token act of driving an SUV one of two things could be true: either (1) this is not the act that will push total emissions over the threshold, thus it is justified despite the impermissibility of the type; or (2) the emissions of this act exceed the threshold only because other SUV drivers have not ceased, or moderated, their SUV driving. (1) questions the causal properties of the token fit to push overall emissions over the threshold per se, and (2) questions the moral relevance of these causal properties, if possessed by the token, given that many other tokens possess identical causal properties, and thus the event of total global emissions exceeding the threshold is causally overdetermined.[70]

Starting with (1), it is true that climate change science does not enable the identification of particular token acts of greenhouse gas emission – indeed, of any such types of act (other than in the broadest possible sense, i.e. 'emitting greenhouse gases') – as causes of climate change. It is likely that climate change science will never enable this. However, it is unreasonable to demand that level of fine-grained causal analysis as a prerequisite for moral requirements and political action to curb greenhouse gas emissions by ceasing certain activities. Indeed, we do not make these demands elsewhere with respect to the regulation of complex and large-scale projects and public health initiatives potentially affecting human welfare, for example the building and location of nuclear power plants, the hazards to public health of pollution from leaded petrol, or the advantages to the long-term health of children of breastfeeding over formula feeding. Daniel Farber makes the general point well:

> If we demand a high level of precision to establish a moral claim, we also render morality irrelevant to the most serious harms created by modern society … Perhaps a nineteenth century court would have required precise matching as a basis for finding liability. But there seems to be no reason why our concepts of justice should remain stuck in the nineteenth century. Nor is it clear why our sense of our own moral accountability should be subject to the same limitations as a court's finding of legal liability.[71]

Turning to (2), Farber addresses the same point made by Posner and Sunstein against holding the US liable (on grounds of negligence) for its failure to curb emissions when other countries fail in a similar way (and indeed, as in the case of China, exceed total US CO_2 emissions).[72] The cumulative emissions objection is that because an agent's emissions create risks of harm only against the background of others' emissions, the agent ought not to be held liable for emissions, because were she to act as her putative liability requires and stop emitting, this would not prevent overall emissions exceeding the relevant threshold. The crux of Posner and Sunstein's argument is exactly this point:

> [I]t is not negligent to fail to contribute to a public good if not enough others are doing similarly, so that the public good would not be created even if one did contribute.[73]

Farber interprets Posner and Sunstein as claiming that there is a 'saturation effect' with respect to CO_2 emissions such that if all emitters other than the agent in question (the SUV driver, or the US) continue to emit, the effect of the agent's continued emissions on climate change will be nil.[74] If there is a saturation effect, then we can interpret Posner and Sunstein as claiming that the US (or the SUV driver) can have no obligation to curb emissions because doing so will make no difference to climate change, given the continued (and rising) emissions of others.

There are some assumptions underlying the 'saturation effect' argument which require a defence. First, and most obvious, is whether there is in fact such an effect. A consequence of the existence of such an effect would be that, beyond a certain threshold at which the effect takes hold, further emissions would have no incremental effect on rises in temperature. As Farber does a good job of showing, there is no evidence that this is the case. In fact, he claims, 'it is possible that Posner and Sunstein have it exactly backwards':[75] according to many models, the damages caused by temperature rises effected by CO_2 emissions may increase in severity as total emissions increase, which could make marginal increases in CO_2 emissions very significant. The uncertainty about how far, exactly, we have gone down the road to irreversible, and perhaps catastrophic, climate change, and the naïveté of thinking that if we have exceeded a safe threshold of emissions, then any further emissions beyond that will not make things worse, makes positing the existence of such thresholds for emissions foolhardy. Given the amount of CO_2 and other greenhouse gases already in the atmosphere, and given the possibility that they could already be having effects that place us on the precipice of climate change tipping-points, perhaps from now on every token act of emitting greenhouse gases creates risks of serious climate change harms. At the very least, assuming that no such token act creates these risks is unsupported by evidence.

A second questionable assumption of Posner and Sunstein's argument is that the obligation to curb emissions exists just in virtue of the actual consequences that the performance of any act discharging that obligation would have. However, as is familiar from debates between consequentialists and their critics, there are other grounds from which obligations can be derived which do not require predicting the effects of actions fit to discharge the obligations. Posner and Sunstein's crude consequentialism is not the only, and by no means the most attractive, way of thinking about how obligations are justified.[76] For example, Kantian approaches have it that a token act is impermissible if, because, and insofar as the maxim according to which it is performed could not serve as a universal law governing the actions of all, and this is determined without consideration of what other people are actually doing, or are likely to do in reaction to the agent refraining (or not) from acting on her maxim. And – more germane to the overall approach I have taken throughout the book – thinking about what is owed to future generations using the Rawlsian device of the original position is a thoroughly non-consequentialist way of determining the demands of climate change justice.

Unless a saturation effect is assumed and a crude consequentialism adopted, the 'cumulative emissions' case does not deliver a justification fit to relieve any emitter of liability. Luxury emissions impose risks of serious climate change harm on future generations and are impermissible regardless of what others are doing, or have done.

Conclusion: an Intergenerational Fund for Climate Change Compensation

To recap what I have argued so far in this long chapter. Any current generation has a duty of justice to future generations not to rob them of the possibility of achieving justice in virtue of imposing unbearable strains of commitment on them. This duty is breached by any generation which emits greenhouse gases so as to put future generations at risk of serious climate change harm. The appropriate principles of justice to address this breach are principles of corrective justice. Responsibility for the risk-imposition is *ex ante*, and as such is justified, given a Rawlsian commitment to equality. The existence of freedom-defeating excuses for subsistence emissions suggests that – at least at this point in time, and given how our energy regimes make these emissions an avoidable necessity – liability for imposition of the risks they create on future generations is fault based. The question now remains how best to ensure that those who may become victims when these risks mature into harms are in a position to get recovery.

There are many ways in which the recovery to which victims are entitled, under corrective justice, can be delivered through institutions. Tort law connects injurers to victims directly, and makes it the responsibility of the victim to pursue the injurer through litigation, and the responsibility of the injurer to provide funds to compensate the victim if litigation succeeds. There are good reasons why intergenerational litigation is not an attractive way to realise intergenerational corrective justice with respect to climate change harms, the most persuasive of which relate (a) to causation, and (b) to ensuring recovery against time's arrow.

Turning first to issues of causation, any litigation by members of a future generation must be for a particular harm, or at least a particular type of harm. Let us take as an example the harm of malarial illness suffered by members of future generations living at latitudes which, for earlier generations, had been free of the disease because the temperatures were too low for mosquitoes to survive and reproduce, but which have become colonised by mosquitoes moving pole-wards as temperatures in these regions increase. In order to establish the liability of earlier generations for the risk of malaria imposed on later generations it would have to be shown that the acts of earlier generations were the cause-in-fact of the migration of mosquitoes. Is it possible to make such precise causal links between the specific greenhouse gas-emitting activities of, say, generation t-6, and the harms suffered by members of generation t? Climate change scientists are universally chary of making any such precise causal claims (witness the widespread reluctance of any expert to attribute Hurricane Katrina to climate change). This creates problems at both ends of the compensation relationship, so to speak. With respect to putatively liable agents, problems with establishing cause-in-fact connections between specific risky acts and particular harms they could ripen into threaten to undermine liability at the defendant end of the relationship. And with respect to putatively harmed victims, these problems threaten to stymie any claims for damages for particular harms suffered in the future at the plaintiff end of the relationship.

Notwithstanding the reluctance of climate change experts to make claims about the chains of causation between the emission of greenhouse gases and particular changes in climate (putatively causing particular weather and other events), an increasing number of lawyers are testing the waters in litigation cases in the US.

For example, in *Comer* v. *Murphy Oil Co.*, fourteen people are suing a collection of oil, coal, and chemical companies for contributing to climate change and thereby increasing the intensity of Hurricane Katrina, as a result of which they suffered losses.[77] The grounds of their suit include negligence, nuisance, and trespass.[78]

However, these cases are intragenerational (and within national boundaries): plaintiffs and defendants are contemporaneous. Herein lie a second set of reasons for eschewing litigation as a means to realise intergenerational corrective justice with respect to climate change harms. At any point in time, any victims suffering harm when the wrongful risks imposed on them by their predecessors mature will be in a position to pursue only a very limited number of injurers, i.e. those still living. For example, members of my generation – Generation X – might legitimately accuse ageing baby boomers of negligence in failing to cut back their luxury emissions, thereby creating some inevitable climate change for us and our children, to which we – and not they – now have to devote resources and time to try to solve. And members of Generation Y who followed my generation probably have an even stronger case against Gen X-ers than we have against the baby boomers. There is always likely to be a limited subset of cases in which persons belonging to a younger generation suffer some harm as a result of risks imposed on them by still living members of an older generation. However, injuries to the younger generation will likely have ripened from risks imposed on them by many dead members of preceding generations, who are beyond reach. Many members of future generations may thus be left without – or with only very limited – means of recovery, if intergenerational litigation is endorsed as the best means for realising intergenerational corrective justice.

A far better proposal is to abandon the idea of litigation in this context altogether in favour of an Intergenerational Climate Change Compensation Fund.[79] As I argued in the section '*Ex ante* responsibility for risk-imposition', the *ex ante* approach to responsibility for emissions makes it possible to establish liability at the point at which emissions are made, and thus possible to impose levies on emitters before the risks they thereby create harm anyone. Gathering the information necessary for the targeted imposition of levies envisaged in the ideal form of the Intergenerational Climate Change Compensation Fund is very hard. The problem is not that of causation that dogs the idea of intergenerational litigation. Inevitably, there will be problems here too: members of future generations applying to the fund will have to show that the harms they suffer have been caused by climate change. But if these are not insurmountable in the intragenerational context – and the emergence of cases such as *Comer* v. *Murphy Oil Co.* suggests they are not – then they should not be insurmountable in the intergenerational context. Rather, the problem relates to identifying those liable for their luxury emissions in order to impose targeted levies on them to sustain the fund, regardless of how subsistence emissions are defined, and the level at which they are set. As already noted, there are easy examples at either end of the spectrum: the day labourer in Kohlua is very likely not liable for his emissions, whereas Bono is liable. But many, many cases lie in between these. How are we to decide the level of levy appropriate for each luxury emitter, or each class thereof? However this question is decided, it is possible that gathering the information required will be extremely costly, and bureaucratically nightmarish, and that for those reasons it is better to simply target levies at the rich, and accept that there may be some in this class – the green hermit millionaire living a wholly self-sustained life – who are unfairly taxed.

A precedent of sorts for a fund so conceived in the intragenerational context is New Zealand's Accident Compensation Corporation, which delivers personal injury cover to residents and visitors.[80] The Accident Compensation Corporation provides compensation for a very wide range of injuries – from death through to physical and mental injury – with a number of causes (accidental, work related, medical treatment, and sexual assault). It extracts levies from employers, employees, the self-employed and motorists to fund different 'accounts' to which those injured can apply for compensation. Application to the Accident Compensation Corporation replaces the pursuit of compensation for personal injury through private litigation. The reason why the Accident Compensation Corporation provides only an 'of sorts' precedent for the type of fund I wish to outline is that it does not single out a particular group as engaging in risky behaviour in imposing its levies: for example, all motorists – not just those who drive over the speed limit – must pay the levy. Thus, the Accident Compensation Corporation scheme functions as an insurance pool rather than a fund for corrective justice. Nevertheless, it is worth bearing the scheme in mind, as it is perhaps the best approximation to a fund for corrective justice in a world in which information gathering is hard.

5

THE END OF THE WORLD AS WE KNOW IT

Catastrophes and triage

Introduction

The focus of this book until now has been on what the current generation ought to do in order to meet its duties of justice to future generations. I shall turn now to the question of what future generations might have to do if (as is likely) the current generation fails in this matter. If we and our children do not change our lifestyles, politics, energy regimes, institutions, and priorities in order to make the drastic reductions in emissions necessary, at best to avert, but more realistically to moderate, the worst effects of dangerous climate change for those who will follow us, what forms of social cooperation (if any) do we bequeath to them as their only options? If we pauperise them not just by ruining the planet, decimating biodiversity, exhausting the atmosphere's limited capacity to absorb greenhouse gases, and by leaving them no fund from which to draw some form of compensation in order to adapt, but also by creating conditions in which – of necessity – 'justice' disappears from their normative vocabulary, then what are the principles, if any, according to which they could live together?

Let us remind ourselves of some sobering facts about the possible effects on the circumstances of future generations of a worst-case business-as-usual scenario. Remember the predictions of the Met Office for temperature rises in this scenario, in which emissions increase 132 per cent over pre-industrial levels by 2050.[1] The worst-case outcome given business-as-usual is a global average temperature rise of 7.1°C (10 per cent chance), and the most likely case is 5.5°C increase (50 per cent chance), by 2100.

Note that these are global averages: average temperature rises in some regions are likely to be far higher, with correspondingly lower rises (or even decreases) in other regions. This means that the effects of climate change on people living in 2100 and beyond will be unevenly distributed; for example, those living in one of the global 'dry belts' of permanent drought predicted by one climate model for a 5°C increase would likely become environmental refugees as local agriculture and food production becomes impossible (Africa and Asia would lose one third of their food supply), whereas those living in Scotland would be unlikely to face this fate (although they would, of course, suffer other consequences, such as a new risk of malarial infection).[2] Other effects of temperature rises above 5°C would likely contribute to the refugee problem in 2100: at least 30 per cent of coastal wetlands would be lost in this scenario, and much of this land is inhabited. The increased risks and incidence of coastal flooding would contribute to the retreat of refugees inland. An increase of 5°C+ would warm the seas and could cause the collapse of ocean methane hydrates, releasing huge amounts of CH_4 into the atmosphere and causing irreversible runaway climate change, punctuated by massive environmental catastrophes, as described in Chapter 3.

It is easy to become numb to what figures such as these mean, and to why they really matter. What sorts of political arrangements would be made by people enduring these conditions? Would nation-states survive, or would political groupings become local and tribal? Or would social and political groupings – outside the family – simply disappear? Would there be any place for, point to, or possibility of international political alliances? Or nongovernmental organisations? And, at the more personal level, how would people in these circumstances try to keep their children safe, and alive? What would happen to those who get old, fall ill, or become frail? To those with limited mobility, visual impairments, or mental health problems? What would life be like for people having to live in the conditions that these temperature increases would create?

Consider the disturbing testimony of Pavlo Makohon, a 15-year-old boy who survived the Ukrainian famine in 1932–33, during which between 2 million and 10 million Ukrainians died:

> In the spring of 1933 famine struck my family. We'd eaten everything in the house and our bodies began to swell. […]
>
> Our village was in the clutches of death. Every dog and cat had been killed and eaten. When the horses began to die in the kolhosp then the villagers ran there in the hope of finding meat. They ran with knives and sacks, and I ran too, and when a horse died we, who were hungry and swollen, attacked it like lunatics, trying to cut off a piece of meat. At first we accomplished this unhindered. But then the authorities invented a way of keeping even these dead carcasses from the starving. They dug a large hole and when the horse was dying the activists poured carbide all over it and threw it into the hole. By now, however, the people were half-crazed with hunger and despair and, ignoring the poison, they scrambled into the hole and hacked at the horse with their knives, running home with their poisoned bounty. It was our last hope of survival, and even this was turned into a death-trap.

> I wandered around the outskirts of the village looking for hedgehogs in the bushes. When there were none left we knew we wouldn't survive. People began to die at an alarming rate. In almost every house the physically-strong succumbed first, then the others followed soon after. The authorities saw that the situation was becoming uncontrollable, for in many homes the corpses of the dead were decomposing. So they organized a team of stronger men to dig a huge hole in the cemetery. To this hole they drove all the dead in carts and flung them in.
>
> Death entered our house. Within a week my grandmother died.
>
> One after another my brothers, Ivan, Vasyl and Anatoly, and then my little sister Maria, also died. They were hauled away in the kolhosp cart to the pit and thrown in. I saw this happen with my own eyes.[3]

The day-to-day priorities for people in a 5°C world would be finding food, water, shelter, and staying safe. Lucky people would be located in habitable regions with access to the (diminishing) resources necessary for survival; the unlucky ones would have to forage daily for these resources, or fight to gain access to a habitable region. Conflict between the 'haves' and 'have-nots' over survival-goods would be inevitable; and the threat of starvation or death from dehydration – either by having resources taken away, or by being denied access to them – would make such conflict relentless and desperate. Furthermore, within the groups of 'haves' and 'have-nots' the weak would be very vulnerable: the sick, the frail and old, and the very young would perhaps have only family to provide them with survival-goods, and would not be able to contribute to any efforts to protect stockpiled resources. In addition to conflict created by scarcity of resources, there is evidence to suggest that increased temperature in itself causes increases in violent aggression: one study predicts a rise of around 24,000 assaults and murders in the USA every year for every increase in average temperature of 2°F.[4] A further layer to the experience of people in this world – not often remarked upon in the literature – is that most of them would have constant exposure to death. A 5°C temperature increase by 2100 would mean that most survivors would have witnessed the death of family, friends, and compatriots from starvation, dehydration, disease, disaster, or conflict: the testimony of Pavlo Makohon would become autobiography for most people. In addition, they would be witnesses to the decimation of the civilisations their grandparents (we) and parents knew, and the swift mass extinction of other species on a scale never before experienced in the history of the planet. The people of the 5°C world would be in a state of grief, anxiety, distress, and panic.[5]

Adequate precautions against climate change are not being taken, no fund exists to compensate future generations for the harm they will suffer as a result of our failures, and there could be no techno-fix adequate to solve the problems of a 5°C world. Here I shall, very tentatively, consider one direction politics could take in the 5°C world by elaborating the content and purpose of principles of triage for the distribution of scarce survival goods in it.[6] Because politics is possible in the absence of states, governments, institutions, and basic structure, it remains possible to talk of politics when these things have disappeared. So long as there are groups of people attempting to live together in some way, political thinking has relevance.

The chapter proceeds as follows. In the next section I identify the levels at which principles of triage can operate, the categories with which they operate, and the goods they distribute. In the section 'Are principles of triage principles of distributive justice?' I consider, and reject, two ways in which to conceive of principles of triage as principles of justice. Notwithstanding this rejection, the section 'Triage and hope for future justice' articulates an important connection between the principles of triage for a 5°C world and intergenerational justice. I argue that the adoption of principles of triage in these circumstances could keep alive hope for a future return to the pursuit of full justice, or at least decency, if and when environmental conditions improve.

Levels, categories and goods of triage

Historically, principles of triage have been adopted to distribute resources in conditions of dire scarcity.[7] These principles sort people into various categories, and state priority rules for the distribution of resources to people in different categories. The classic contexts for triage are medical treatments involving scarce goods (e.g. donor organs); battlefield medicine, war-time rationing, and disaster/emergency medicine and rationing. In many of these scenarios principles of triage are an application or interpretation of a background theory of justice in unusual, often temporary, nonideal circumstances. (For example, with respect to battlefield medicine, the conduct of enlisted physicians must conform to principles of *jus in bello*.)

Principles of triage can serve for the micro-allocation of resources, the macro-allocation of resources, or both. In the former case, they guide the decision making of on-the-ground personnel charged with distributing scarce goods, often under great pressure. Here, principles of triage often operate at best as rules of thumb for people who may have just seconds to sort people into categories and distribute resources according to the priority rule in use, such as in the aftermath of a natural disaster, or on the battlefield. However, triage at the macro-allocation level is also possible, and can be appropriate, when personnel have longer to assess persons for the purposes of categorisation, and to apply the priority rules; for example, principles of triage may operate in transplant assessment centres to which patients are sent for a period of days prior to being placed on a waiting list (or not).

In most cases of triage at the micro-allocation level, the principles employed are either an interpretation of principles of justice which operate to allocate resources at the macro-level (the principles for micro-allocation are an attempt to apply otherwise justified principles of distributive justice in the limited domain of scarcity of a particular good needed only by a small and unlucky proportion of the population), or are principles of local justice which may have a less obvious connection with principles of justice at the macro-level.[8] However, there is no necessity in this: triage at the micro-level can be an interpretation of triage at the macro-level, where the macro-allocation makes no claims to interpret justice, or it can stand freely as an expression of the initiative of personnel on the ground in the absence of any principles of triage at the macro-level.

Traditionally, those in need of goods and resources to be triaged have been sorted into three crude categories which reflect the nature of the goods distributed by principles of triage as necessary for saving life, and the circumstances in which triage is appropriate and required, i.e. conditions in which there are insufficient resources to save the life of each person. These categories are:

A. Those likely to survive without any resources.
B. Those likely to die even with resources.
C. Those likely to survive only if they receive resources.

All systems of triage share a commitment to prioritising category C over A and B, for reasons of efficiency: when resources are very scarce, to direct them to categories A or B is likely to be a waste: A-people already meet what I shall call the 'survival likelihood' criterion, and B-people are unlikely to do so, whatever we do.

In addition to a 'survival likelihood' criterion, a 'survival mutability' criterion is also often used in triage. This criterion sorts people in each category according to whether they are likely to survive for a reasonable period of time with, or without, resources.[9] Note that the survival mutability criterion applies only to categories A and C, because people in B are unlikely to survive at all. So, the categories can now be further refined as follows:

A1. Those likely to survive only for a short period of time without any resources.
A2. Those likely to survive for a reasonable period of time without any resources.
C1. Those likely to survive only for a short period of time only if they receive resources.
C2. Those likely to survive for a reasonable period of time only if they receive resources.

A third criterion, sometimes applied, relates to quality of life, and applies (again) just to categories A and C (because B-people are likely to have no life to which the criterion could apply). In the practice of triage, this criterion introduces the thought that scarce resources ought not to be used to save those who meet the other criteria but who are likely to persist in a vegetative state. It is possible that the 'quality of life' qualification could be stretched beyond these types of cases to include people whose lives can be saved with resources, but who have a condition that will not be improved by the resources and will rob them of all quality of life, despite not being in a persistent vegetative state. An example might be of a starving person who is likely to survive if given food, but who has a non-fatal but excruciatingly painful permanent medical condition requiring morphine for its control. If food, albeit scarce, is available for distribution through triage, but there is no morphine available at all, then the 'quality of life' qualifier could operate to place such persons in category B rather than C. I do not make this claim here, but it is a possibility, and might become an increasingly plausible option if scarcity widens, deepens, and entrenches for people experiencing disastrous runaway climate change.

Overlaying the 'quality of life' criterion onto the other criteria gives the following categories:

A1a. Those likely to survive only for a short period of time without any quality of life without any resources.
A1b. Those likely to survive only for a short period of time with quality of life without any resources.
A1c. Those likely to survive for a reasonable period of time without any quality of life without any resources.
A1d. Those likely to survive for a reasonable period of time with quality of life without any resources.
C1a. Those likely to survive only for a short period of time without any quality of life only if they receive resources.
C1b. Those likely to survive only for a short period of time with quality of life only if they receive resources.
C1c. Those likely to survive for a reasonable period of time without any quality of life only if they receive resources.
C1d. Those likely to survive for a reasonable period of time with quality of life only if they receive resources.

When dire scarcity contains these categories, triage becomes far more complicated. Retaining the assumption that only C-people will benefit from the resources available for triage, it is plausible that C1d would take priority over the other sub-categories within C, and that C1b would take priority over C1c and C1a. (Of course, this ordering could be questioned: if what matters is extending life per se, regardless of its quality, the C1c will take priority over C1a and C1b.) However, it is not clear which of the remaining two categories – C1c and C1a – would be next in line for resources: the former is more likely to survive for a reasonable period of time than the latter, given resources, but both are likely to lack any quality of life. If the 'quality of life' criterion is held to be very significant (and especially if the lack of it in the cases in hand is acute, e.g., constant and excruciating pain, for which no medication is available), then it may be that the practice of triage relocates these categories to the group of people (already comprising A1a–A1d, and all of B) to whom no resources ought to be distributed. This relocation would be tantamount to 'passive' euthanasia, whereby the mercy killing is performed by withholding the resources essential for survival. However, unlike euthanasia in 'normal' contexts wherein the killing is not made necessary because of a dire scarcity of resources but rather responds to the request of the subject or their family, this form of killing might sometimes (perhaps often) be performed without the consent of the subject, for want of resources. In circumstances of dire scarcity such as runaway climate change could create for our grandchildren, non-voluntary – and perhaps involuntary – euthanasia could become a common practice.

Because those formulating principles of triage for macro-allocation, and those practising triage at the micro-level, very often do not know for sure that a person belongs in one category rather than another, sorting is done in terms of probability judgements. Of course, this inevitably leaves room for tragic errors when a person is sorted into category A or B but should in fact have been sorted into C, and dies as a result. Given the absence of crystal balls, and always assuming that best practice for sorting is applied, there is no way to avoid such errors completely.

Next, it is important to be clear about the types of goods that triage distributes. Clearly, there can be dire scarcity of almost any type of good, but not all goods are open to distribution, and some goods – by their very nature – are not fit for distribution by principles of triage; for example, social and personal goods such as the goods of relationships, association, self-respect, play, etc. Although other resources can certainly be distributed so as to make it more easy or difficult for people to achieve these goods, they cannot themselves be directly distributed. Indeed, any attempt to do so is likely to destroy the good in question: a person assigned to me by the Directorate of Friendship is not *ipso facto* my friend (although our subsequent free interaction may lead us to become friends). Social and personal goods are essential to a good life, and a dire scarcity of material goods would destroy many important opportunities for the enjoyment of these goods. But the nature of them means that principles of triage cannot address the scarcity of them.

So what are the goods that triage addresses? These are survival goods: food, water, shelter, medical care and treatment, and security. With the possible exception of the last type, these goods are material and tangible, and have a history of distribution by principles of triage. It may even be the case that greenhouse gas emissions themselves are presently a survival good, because existing energy regimes make it impossible to survive without engaging in activity which necessitates the emission of some CO_2 (or equivalent); that is, we may already be in a scenario of dire scarcity vis-à-vis this good. As Henry Shue suggests, if per capita emissions available (given the global limit for CO_2 emission necessary to avoid runaway climate change) are less than per capita survival requirements (given the current energy regime), then we already have a dire scarcity of a survival good.[10]

The 5°C world is a set of circumstances in which principles of triage are appropriate for the distribution of survival goods insofar as it contains groups of people with access to direly scarce resources, and capable of using the principles to guide their macro- and micro-allocation decisions. However, it might be that the unlucky people in the 'have not' category cannot be said to constitute such a group. Instead, they might be a collection of scavenging individuals, too scared and desperate to coordinate their actions so as to do something together. In that case principles of triage are possible and appropriate only for the lucky 'haves', whose access to survival goods is likely to make possible some degree of coordinated action.

Are principles of triage principles of distributive justice?

Principles of triage express considerations of efficiency with respect to the distribution of survival goods in conditions of dire scarcity. Are they *ipso facto* also principles of distributive justice? It might be thought that the answer to this question does not matter – and certainly would not matter to people in the 5°C world attempting to implement the principles. However, as I shall argue in the section 'Triage and hope for future justice', the status of these principles does matter from the point of view of understanding how 5°C-people could be members of an intergenerational social union committed to the achievement of justice. This point of view is likely to have no resonance whatsoever for these people, but it does have resonance for us, and it is important to see that there might be grounds for hope for humanity even in the bleak 5°C world.

For now, my question is: would the 5°C world provide circumstances in which distributive justice is possible?[11] This question can be expanded in at least two ways, each of which expresses a different perspective on the circumstances of a 5°C world.[12] First, we might be asking whether justice can be predicated of the circumstances at all: are the circumstances such that relations between people in them could be (distributively) just or unjust? Do the circumstances furnish necessary or sufficient 'existence conditions' for distributive justice?[13] Or, we might be asking whether distributive justice is feasible in the circumstances: if we can rightly say that there is injustice in the circumstances, is it practicable to make the transformation to justice? Do the circumstances furnish 'feasibility conditions' for justice? Debates about the feasibility of justice in various contexts are often speculative and underpinned by incompatible visions of human nature and the possibilities for institutional design. However, if the existence conditions for justice are not supplied by a set of circumstances, then questions about feasibility do not arise. So, my focus will be whether the 5°C world would provide existence conditions for distributive justice. I shall argue that the 5°C world would fail to provide such conditions with respect to two prominent conceptions of distributive justice, and thus questions about the feasibility of distributive justice do not arise. These conceptions are framed in terms of (1) pure procedural justice, and (2) the allocation of goods and services. The upshot of this discussion will be that the language of justice is misplaced in application to the distributions of good effected by principles of triage: triage is not a form of justice.

Before beginning, it is worth revisiting Hume's famous description of why the existence conditions for distributive justice would be absent in circumstances of dire scarcity.

> Suppose a society to fall into such want of all common necessaries, that the utmost frugality and industry cannot preserve the greater number from perishing, and the whole from extreme misery; it will readily, I believe, be admitted, that the strict laws of justice are suspended, in such a pressing emergence, and give place to the stronger motives of necessity and self-preservation.[14]

In the Met Office's very worst-case scenario of a 7.1°C temperature increase by 2100 the goods and resources necessary for survival are so scarce that there might be insufficient for any to survive, i.e. the species could go extinct. This is dire scarcity indeed, and the extinction of *Homo sapiens* clearly does not provide an existence condition for distributive justice. Putting aside this possibility, there are two features of the (bad enough) 5°C world relevant to the assessment of whether it would provide existence conditions for distributive justice. First, at best there would not be enough for each person to receive the goods enabling them to survive. And second, when goods are so scarce as to threaten the survival of 'the greater number', it is likely – indeed, acceptable – that 'the stronger motives of necessity and self-preservation' will dominate the behaviour both of the people who are threatened, and of the people who are not yet threatened but who know that they soon could be.[15]

Let me start with consideration of the first conception of distributive justice, framed in terms of pure procedural justice. The question to which principles of pure procedural justice provide an answer is, in the words of its most famous advocate, Rawls:

> [H]ow are the institutions of the basic structure to be regulated as one unified scheme of institutions so that a fair, efficient, and productive system of social cooperation can be maintained over time, from one generation to the next?[16]

For Rawls, a necessary existence condition for distributive justice is social cooperation, which has 'at least three essential features':

(a) 'Social cooperation is distinct from merely socially coordinated activity … [it] is guided by publicly recognized rules and procedures which those cooperating accept as appropriate to regulate their conduct';
(b) 'The idea of cooperation includes the idea of fair terms of cooperation [which] each participant may reasonably accept, and sometimes should accept, provided that everyone else likewise accepts them. Fair terms of cooperation specify an idea of reciprocity or mutuality: all who do their part as the

recognized rules require are to benefit as specified by a public and agreed upon standard';

(c) 'The idea of cooperation also includes the idea of each participant's rational advantage, or good. The idea of rational advantage specifies what it is that those engaged in cooperation are seeking to advance from the standpoint of their own good.'[17]

Each of these conditions is necessary for social cooperation (and they may be jointly sufficient). Social cooperation is a very specific form of joint activity: it is governed by rules accepted as appropriate by those engaged in it, where these rules distribute benefits to participants according to a standard of reciprocity, and where these benefits at least partly advance the participants' rational advantage. It is not the case that any form of joint activity qualifies as cooperation. The purpose of Rawls' pure procedural theory of distributive justice – justice as fairness – is to provide principles to govern the design of a basic structure (comprised of the 'main political and social institutions of society')[18] qua framework for just social cooperation.

Furthermore, Rawls argues, only the basic structure can provide this framework, for at least the following reason: the background conditions against which individual transactions are fair have a tendency to erode as a result of the cumulative consequences of individual choices and transactions which may, in themselves, be acceptable. For example, voluntary market transactions made freely can result – as Nozick's 'Wilt Chamberlain' example shows – in a distribution of income and wealth that undermines the fairness of these transactions by skewing the distribution of 'fair opportunities to earn income, to learn wanted skills, and so on'.[19] As Rawls puts it,

> 'The role of the institutions that belong to the basic structure is to secure just background conditions against which the actions of individuals and associations take place. Unless this structure is appropriately regulated and adjusted, an initially just social process will eventually cease to be just, however free and fair particular transactions may look when viewed by themselves.'[20]

Although pure procedural justice is a property exclusive to a just basic structure, and although the purpose of a just basic structure is to provide a framework for just social cooperation, it is not the case that if social cooperation is absent, this conception of justice has no place. If this were the case, then the Rawlsian vision of justice would be peculiarly reactionary and conservative: it would issue no injunctions that unjust forms of joint activity be transformed into just forms of social cooperation, through the construction of a basic structure the institutions of which frame this cooperation to provide a background of justice, making pure procedural justice a reality. As Arash Abizadeh puts it:

> [U]sing the ideal of social cooperation to specify (and restrict) the scope of application of the ideal itself … perversely implies that demands of distributive justice arise only between persons whose social interactions are *already* conducted on fair terms, i.e., that demands of justice would not arise for persons whose social interactions are unjust.[21]

Abizadeh plausibly claims that the scope of the ideal of pure procedural justice is fixed not by reference to existing social cooperation per se, but rather by reference to 'the range of persons engaged in a system of social interaction … that is or *could become* a scheme of social cooperation. If a system of social interaction does exist, and if a just basic structure could exist, then there ought to be a basic structure to which the principles of justice can be applied.'[22] In sum, a necessary existence condition for pure procedural distributive justice is disjunctive: there must be present *either* social cooperation that could be governed by a just basic structure, *or* social interaction that could become such social cooperation.

To bring this discussion to bear on the triage of survival goods in the 5°C world, in a situation in which 'the stronger motives of necessity and self-preservation' dominate human behaviour, social interaction which is, or could become, just social cooperation will be unlikely to be present. Bearing in mind the immediacy of the day-to-day priorities of hungry and scared people, we can see this by considering whether social interaction in the 5°C world could meet Rawls' conditions (a)–(c) for social cooperation (laid out on pp. 115–116). With respect to conditions (a) and (b), it is possible that in some groupings – perhaps families, or those with relatively secure access to limited resources – cooperation could be governed by publicly recognised rules of triage, and that all could accept that everyone who does what is required by the principles of triage deserves to benefit as stipulated by the principles. This could involve, for example, submitting to medical examinations and assessments, waiting one's turn in the assessment queue, not using more resources than required for one's survival, or accepting a decision that principles of triage dictate that others' needs are more urgent than one's own, or those of one's family. However, it is not so clear that social interaction in the 5°C world could meet condition (c), that 'cooperation … includes the idea of each participant's rational advantage, or good'. The dire scarcity in the 5°C world means that not all participants in a system of social interaction governed by principles of triage advance their rational advantage, even partly. Direly scarce goods to be triaged in a 5°C world are survival goods, and thus any system of social cooperation to effect a triaged distribution of these goods will have as a direct and unavoidable outcome the death of some participants who could have lived, had more of these goods been available. If there is anything which is not a component of a person's rational advantage, even partly, it is their own death. This fact disqualifies principles of triage as principles of justice because the unavoidable outcome of the principles makes them unjustifiable in the name of justice to those affected by them. When a person knows that principles adopted to govern her social interaction with others place her at direct risk of death – if not now, then maybe tomorrow, or the day after – then her objection to the principles on the grounds that they do not deliver justice to her (or her loved ones) is not unreasonable.

Note here that the claim being made is not that such principles are unjustifiable per se, or even that the justification of them cannot be made by reference to some ideal of justice (I shall say more about this in the next section). The claim being made is just that principles of triage for such circumstances are not principles of *distributive* justice, when the latter is conceived in pure procedural terms. That is, we cannot say of a triaged distribution of goods which has as a direct result the death of some who could have been saved had more resources been available that it achieves distributive justice.

However – and to turn to the second conception of distributive justice to be addressed – if distributive justice is conceived of in terms of the *allocation* of goods to persons, rather than in terms of pure procedures the outcomes of which are guaranteed to be just, then principles of triage appear more plausible as principles of justice. Allocative and pure procedural justice differ with respect to their subject matter: pure procedural justice is concerned with social cooperation within the framework of a basic structure, whereas allocative justice is concerned with the distribution of goods to persons whether or not this is effected through the institutions of a basic structure. Nevertheless, they are part of the same family of conceptions of distributive justice because their purpose is to ensure that the share of goods received by persons in a society governed by them is just.

Allocative justice consists of principles to determine how goods ought to be distributed to persons just in virtue of how some feature of the good in question relates to some feature of the recipient. For example, a principle of allocative justice might distribute goods to persons according to their merit, according to how the good would improve their level of well-being, or according to their needs. In contrast, principles of pure procedural justice determine how institutions governing social cooperation ought to be designed in order to secure a background against which the sum of the outcomes of persons' choices is *ipso facto* a just distribution of goods.[23] Of course, principles of pure procedural justice also respond to features of persons, who are in receipt of justice in virtue of features they share. However, and *pace* the allocative view, these features are not treated as markers specific to particular individuals in virtue of which a particular good, or category of good, ought to be distributed directly to them. The question to which principles of allocative justice provide an answer is: who gets what? The context for allocative justice is: a set of persons with competing claims to a divisible good, some distribution of which does, or could, satisfy the standards of the principles of allocative justice. In contrast, the context for pure procedural justice is: social cooperation producing benefits and burdens within the framework of a basic structure, or social interaction that could become cooperation so described. The context for the latter is absent in the 5°C world, but the context for the former might not be.

If distributive justice is allocative, does the 5°C world provide conditions in which principles of triage can stand as principles of distributive justice? From the perspective of allocative justice, the fact that any triaged distribution of survival goods in dire scarcity has as an unavoidable outcome the death of some does not per se disqualify principles of triage as principles of allocative justice; rather, it simply shows (at best) that principles of justice so conceived are not feasible. Although it might be desirable (perhaps for reasons of stability) that principles of allocative justice are supported by reasons no one affected by them could reasonably reject, this is not necessary for these principles to be justified per se. If the outcome of a triaged distribution of direly scarce survival goods is the death of some, the fact that those who will die can reasonably reject the principles according to which the distribution is made – even when there are no alternatives – does not *ipso facto* show that those principles are not just; rather, it shows that it is not feasible to expect people to abide by them. It may be that the principles are just for reasons entirely unrelated to the considerations that make them feasible; or it may be that they are unjust even though feasible. On this view, the badness of a distributive outcome from the first-person perspective (and outcomes do not get much worse than death) is insufficient to show that outcome to be unjust, whereas according to the pure procedural view, there is a threshold of badness for persons in outcomes below which an outcome is not just, in virtue of how the first-person perspective on that outcome makes principles leading to it unjustifiable (and death is certainly below this threshold).

If principles of triage are principles of allocative justice, then it must be claimed of triaged distributions of goods in virtue of which some are condemned to death that this outcome for these people is just, despite their (doubtless vehement) rejections of it as such. Principles of allocative justice distribute goods to persons in virtue of their possession of some relevant property according to which they have a valid claim to the good. Principles of triage distribute survival goods by using priority rules – framed in terms of survival likelihood, survival mutability, and quality of life – related to the nature of survival goods. These criteria appeal to considerations of efficiency: they order people in terms of their capacities to be efficient converters of resources into life of a reasonable length and some quality. In circumstances of dire scarcity the use of such efficiency considerations is probably the only alternative to an outright war of all against all, in which the strongest and most ruthless gain possession of resources without which they could have survived, causing the death of others whose survival depends on these resources. When these are the alternatives, principles of triage qua principles of efficiency appear reasonable and attractive. However, that principles of triage are the only reasonable option in a set of circumstances does not show that they achieve justice in those circumstances. To claim this, an advocate of the allocative view would have to argue that a person who receives a triaged share of resources is deserving of – has a right to – them, which entails the claim that a person who does not receive resources does not deserve – and has no right to – them, even when this person dies as a direct result of being refused resources. However, the property 'being an efficient converter of resources into life' lacks the moral significance necessary to support these claims about just desert. Possession of this property in most circumstances is a matter of brute luck: a person's genetic inheritance, her age, her upbringing, her place of birth and living, her diet, her level of education, her exposure to natural disasters and toxins in her environment, etc. all bear on the extent to which she is capable of converting resources into life.[24] And this is especially the case in the 5°C world, wherein those weakened by disease, malnutrition, and dehydration brought on by conditions of accelerated climate change cannot be said to be responsible for the environmental disasters that threaten their lives.

In sum, for principles of triage to be defended as principles of allocative justice requires a defence of the moral significance of possession of the property 'being an efficient converter of resources into life' fit to justify the outcome of death for those who lack it, or have less of it than others. I submit that no such account is available. Given this, and the argument made earlier with respect to pure procedural justice, it can be concluded that principles of triage are not principles of distributive justice, at least with respect to the two dominant conceptions considered here.

An objection to triage in the face of this conclusion might be that, if principles of triage are not principles of justice, then we ought not to adopt them at all, despite their status as principles of efficiency. The thought behind this objection might be that nothing less than justice will do for social organisation, and certainly nothing less than justice can justify forms of interaction which condemn some to death. On this view, it might be claimed that the following principle is preferable:

> *The Levelling-Down principle*: none should be saved if not all can be saved.[25]

There are at least three things to note in response. First, the Levelling-Down principle remains a principle of triage, given the circumstances of dire scarcity in which it is proposed that it operates: it is simply an inefficient principle.

Second, if the efficiency considerations informing the principle of triage addressed so far are unjustifiable in the name of justice insofar as they condemn some to death, and if these are the grounds on which the principle is rejected, then the Levelling-Down principle must be defended qua principle of justice despite the fact that it would condemn all to death, even though some people could be – perhaps relatively easily – saved. This is not to say that the efficiency-based principle is preferable just because, according to it, more people would be saved than were the Levelling-Down principle to be used. That is, it is not to say that the decisive factor in justification of principles of triage is the number of people saved: indeed, there are other reasons, not related to numbers per se, according to which principles of triage are justified (as I discuss in the next section). Perhaps there is some conception of equality informing the Levelling-Down principle according to which this principle stands as a principle of justice in circumstances of moderate – and non-life-threatening – scarcity.[26] But it is hard to see how such considerations are morally robust enough to support the claim that the Levelling-Down principle, qua principle of *triage*, is a principle of justice in circumstances of dire scarcity.

Third, the Levelling-Down principle is likely to be even less feasible than efficiency-based principles, given its outcome of death for all. Any attempts to implement this principle would likely be stymied by the strongest and most ruthless people simply taking the resources they want, or which are needed by their loved ones.

Triage and hope for future justice

That principles of triage are not principles of justice does not establish that they are not justified in the circumstances of the 5°C world; but the considerations according to which they are justified need not be limited just to those of efficiency. In this section I shall argue that certain principles of triage could be justified in the dire scarcity of the 5°C world for reasons that go beyond efficiency and are related to how the principled nature of joint activity to achieve triage in desperate times provides grounds for hope for a return to principled joint activity to achieve justice in the future. Hope is likely to be in short supply in the 5°C world: its cultivation is more than just a matter of psychological or emotional well-being, and makes an important contribution to the possibility of justice. Given this fact, I shall argue that the connection between principles of triage and hope provides some important moral reasons as to why 5°C people ought to practise triage, which complement the efficiency considerations laid out above in 'Levels, categories and goods of triage'.

In order to consider how the practice of triage could cultivate hope, more specification of what triage can and ought to require is necessary. The account of triage above framed its categories and priority rules in terms of the efficient conversion of resources into life of a reasonable length with some quality. However, these rules will not settle all cases, especially as resources diminish and numbers of people in need increase. I shall consider the hard case of when claims compete; that is, when not all those who could be saved for a reasonable period of time with quality of life can be saved with the resources available. Call this the 'competing claims' problem.[27] To solve this problem, practitioners of triage will have to appeal to considerations beyond those of efficiency: the categories and priority rules of triage as outlined so far do not permit distinctions fine grained enough to address the problem.

Although distributive justice is not achievable in the 5°C world, it remains possible that the principles according to which people organise themselves could express commitment to values related to justice, in particular, to the pursuit of justice as an intergenerational project. Here, I shall consider an approach to the competing claims problem and evaluate it in terms of the extent to which it could encourage the cultivation of hope for justice among people unlikely to have any memory of forms of social organisation which could make justice possible, or any expectation that they or their descendants will experience such forms of social organisation for themselves, perhaps for centuries or longer. It is important to be clear about what I shall be claiming: I shall not be arguing that the approach I consider ought to be adopted in preference to all others by people of the 5°C world. The conditions in this world are so far from anything that we (relatively) lucky people have experience of that to argue definitively that a particular approach is, from a moral perspective, the one that 5°C people are required to adopt would be a mistake. There are a variety of principles fit to solve the 'competing claims' problem, and for circumstances in which the world as we know it has ended, it is unwise to argue with authority that any one of these is *the* best approach for people seeking to cultivate hope in hopeless times. The people best placed to judge this are those struggling to survive in these times. I shall give an account of why the approach I shall describe provides grounds for hope for future justice – and thereby make the case for including it in the set of approaches to be considered by 5°C people – but it may be that other approaches prove to be better for people facing a future more bleak than our imaginations can picture. The approach I consider illustrates a way in which to satisfy a moral requirement under which 5°C people stand: to cultivate hope for a future return to the pursuit of justice. My aim here is simply to show how this requirement can be used in arguments for principles of triage. Let me begin by saying more about hope and the requirement to cultivate it with respect to justice.[28]

Hope aims at an objective which exists in the future, is believed to be good by the hoper, and is desired by the hoper in virtue of this belief. Hope generates a disposition to act so as to make the realisation of hope's objective more probable whenever possible, all else being equal. Without this disposition, a hoper lacks the practical commitment to her objective that is characteristic of hope: we would think it odd to describe a person as hoping for an objective if she fails to act so as to realise the objective when presented with a real opportunity to do so.

Hope involves a belief about the future in which its objective exists. This belief can be characterised in two ways: (1) that the objective is possible; (2) that the objective is probable. On the 'possibility' interpretation, hope involves the belief that the objective of hope is both logically and physically possible. Commitment to the logical possibility of hope's objective provides a minimal constraint on the content of the belief about the future in which hope's objective exists: all it rules out is hope for an objective which contains a formal contradiction. Commitment to the physical possibility of hope's objective is a more demanding constraint. What I mean by the claim that an objective is 'physically possible' is that, given what we know about the world and the agents inhabiting it, that objective could exist in the world.[29] Thus, commitment to this constraint ensures that the hoper believes that her objective could be realised in the world she inhabits. Without this constraint a person could counter-intuitively be said to hope for an objective she believes to be logically possible, but which she believes could not be realised in the actual world. To illustrate, consider a person who believes that it would be a good thing to live forever, desires this for herself in virtue of this belief, and is disposed to do things she believes will increase the odds of her living forever, whenever possible, even though she knows that, qua human, she cannot live forever. On the account of hope just suggested, this person is not accurately described as hoping to live forever. Although the proposition towards which her attitudes of belief and desire are directed – 'that I should live forever' – involve no logical contradiction, the state described by this proposition – that of eternal life – is not physically possible, given human biology. Rather than describing this person as hoping for everlasting life, we would describe her as wishing for or fantasising about it.

On the second 'probability' interpretation, hope involves not only the belief that the objective of hope is possible, but furthermore the belief that it is probable. This interpretation should be rejected on the grounds that it has the unwelcome implication that hope collapses either into blind faith or into optimism, and there is good reason to think that hope is distinct from both these attitudes towards future objectives. Let me explain.

There are two ways in which the judgement that hope's objective is probable could be supported. First, by reference to the hoper's belief that she has evidence to support this probability judgement, and second, by reference to a non-evidence-related belief held by the hoper (for example, a belief in divine providence). If the hoper's judgement that her objective is probable is supported by an evidence-related belief, then her hope takes the form of optimism.[30] The optimist believes that the future is likely to bring to pass those things she desires and believes to be good; the optimist believes, let us say, that the probability of the objectives she desires is 0.5 or more. The optimist's reason for believing this is that she believes she has evidence that it is probable that her desires will be fulfilled. The evidence that the optimist marshals to support her judgements about the future sometimes turns out to be nothing more than a projection of her own good will onto the world: optimists are often nothing but wishful thinkers. However, that is beside the point. The optimist believes that she has evidence for judgements about the future, regardless of whether the evidence she believes herself to have is good evidence (or evidence at all): in virtue of the evidence she believes that she has, the optimist thinks that the future is laid out in a way that is likely to satisfy her desires for the objectives about which she is optimistic. In contrast, if the hoper's judgement that her objective is probable is supported by a non-evidence-related belief, then her hope takes the form of blind faith. Here, the hoper judges that the probability of her objective is 0.5 or more, and supports this judgement by reference to, for example, her belief in the existence of a benevolent god.

Refraining from judging the probability of an objective is characteristic of hope for any objective: hope is characterised by uncertainty with respect to its objectives, which shows that it is a mistake to characterise it in terms of probability judgements about these objectives. To have hope is not to 'pocket the future in advance'.[31] Of course, hope can be – and often is – accompanied by blind faith or optimistic judgements about the probability of hope's objective. But these judgements are not components of hope.

This fact about hope explains why pessimism with respect to an objective differs from hopelessness with respect to that objective. Pessimism with respect to a specific objective, like optimism, is either premised on the pessimist's belief that she has evidence to support her belief that the objective about which she is pessimistic is improbable, or on a non-evidence-related belief (perhaps the pessimist is just a misery guts, or believes in the existence of a malevolent god). In contrast, a loss of hope with respect to an objective does not depend on judgements about the objective's improbability; indeed, it is a feature of hope that it often becomes more intense the less probable the objective is believed by the hoper to be. In that case, pessimism about an objective and hope for that objective are consistent: hope can be retained even in the bleakest of circumstances. A loss of hope only attends the hoper's judgement that her objective is impossible, or contra-certain. Upon making such a judgement the hoper despairs of realising her objective.

To relate these reflections on hope to the 5°C world, it is likely that the scared, hungry, and desperate people of this world will be pessimistic about the prospects for the emergence of social cooperation fit to support the pursuit of justice in the dire circumstances of their lives. That is to say, 5°C people will likely judge it improbable that the pursuit of justice will come to animate their joint activities, or those of their immediate descendants. However, as we have seen, this does not show that hope for the pursuit of justice is unavailable to them. In all likelihood, they will in fact be in a state of despair with respect to this objective, but there is no necessity in this. Hope for justice requires the judgement that justice is possible, that it is good, a desire for it in virtue of the belief that it is good, and a disposition to act so as to bring it about whenever possible: these states are available to 5°C people, and there are forms of social interaction between them fit to encourage (or hinder) the development of these states. Indeed, if hope for justice is morally required – which means that the beliefs, desires and dispositions constitutive of it are morally required – then forms of social interaction fit to support the development of these states are also required.

Is hope for justice morally required? We can begin to argue the case by invoking the principle '"ought" implies "can"'. This principle states that it must be possible to perform the actions, or create the state of affairs, demanded by any moral requirements that can be expressed in an 'ought' statement: we cannot be required to do that which it is impossible for us to do. Prima facie, if the principle is true, then if justice is morally required, hope for justice is morally required, because hope for an objective involves the belief that that objective is possible.

This argument fails to establish that hope is required because the belief that an objective is possible does not establish that it is possible, and it is the latter property of an objective that the principle states as a precondition for the existence of a moral requirement to realise that objective. However, the argument can be made to support the requirement to *cultivate the belief that justice is possible* by reflection on *what it takes to make justice possible.*

In the 5°C world there are facts about dire scarcity, and how it affects people's behaviour, which make the pursuit of justice impossible. However, the impossibility of justice here is physical, not logical, and the circumstances of the 5°C world mean that this impossibility is not permanent.[32] There will be many things beyond the control of 5°C people which affect the physical possibility of justice. But there are some things that are within their control; in particular, their attitudes towards the ideal of justice. Facts about the 5°C world make the ideal of justice physically impossible therein. But the permanence of this impossibility is also affected by the beliefs of people in this world about the possibility of others' pursuit of the ideal in improved circumstances in the future. When the possibility of an objective is denied by a person, then her capacity to act so as to satisfy the requirement to realise the objective is impaired, and if this denial is deep rooted, repeatedly affirmed, and never questioned it may become a self-fulfilling prophecy: that which is repeatedly asserted to be impossible thereby becomes impossible.[33] If this is true, then the requirement to realise the ideal of justice entails – in virtue of the principle '"ought" implies "can"' – the requirement to believe that justice is possible. Given other properties of the objectives constitutive of the ideal of justice (that they are good, and ought to be desired as such, yielding a disposition to act so as to realise them), there is a requirement to hope for justice.[34]

One good way in which to cultivate a belief that something is possible in the face of great temptations to despair of it as impossible is to act as if the thing is possible.[35] Here we find the root of considerations favouring some families of solutions to the 'competing claims' problem over others: there are solutions to this problem which, while not themselves informed by principles of justice, are nevertheless better than other solutions in terms of enabling people to act *as if* justice is possible, and thereby fulfil the moral requirement to hope for justice, if not for themselves, then for future generations.

Returning to the 'competing claims' problem, then, let me outline a simple procedure for its solution which would be available to people in the 5°C world, and which would provide them with grounds for hope for justice. The 'competing claims' problem arises when goods are so scarce that the claims of all those in a given category cannot be met, and efficiency considerations can do no further work in terms of prioritising these claims. In such cases of conflict, triage could proceed according to the Equal Chance principle:[36] each person with a valid claim ought to be given an equal chance of having that claim met. When the claims in conflict are numerous, a simple lottery could be used to implement the principle: those with numbered tickets below a threshold number (corresponding to the number of people who can be saved with the resources available) will receive resources, those with numbered tickets above that threshold will be denied them.

There are at least two reasons why, in the circumstances of the 5°C world, use of the Equal Chance principle to solve 'competing claims' problems could provide 5°C people with grounds for hope for future justice. First, the principle respects the separateness of persons:[37] each *individual* is issued a numbered ticket and entered into the lottery as such. Second, the principle embodies impartiality insofar as each entrant is issued a ticket just on the grounds of their having a valid claim to the scarce resources available: sex, race, religion, class, etc. have no bearing on the distribution of tickets and are not taken into consideration.

An objection here is that the lottery procedure makes the receipt (or not) of resources entirely a matter of chance. In the same way that 'being an efficient converter of resources into life' lacks sufficient moral significance to support the outcome of triaged distributions of resources as just, so 'having a winning ticket' similarly lacks such significance. When resources are triaged according to persons' possession of this property, such that those who lack it die as a direct result of the procedure, the principle that informs the procedure is not a principle of justice. However, no such claim is being made for this principle. In the context under discussion people will die whatever we do, and the Equal Chance principle is less unacceptable than alternatives such as the Levelling-Down principle.

For what reasons, then, is the Equal Chance principle less unacceptable than the alternatives? Two of the features of the principle identified above – its impartiality and its individualism – are distinctive features, certainly of the Rawlsian conception of justice, but also of liberal approaches to justice in general. Of course, 'justice' is not synonymous with 'liberalism', and there may be (for example) conceptions of justice not premised on individualism towards which the hopes of 5°C people may better be directed: many varieties of utilitarianism have this character. If this is the case, then perhaps a principle other than the Equal Chance principle would provide a better solution to the 'competing claims' problem. I leave this question open. My aim in the preceding discussion has been far more modest, namely, to show that one dimension of value relevant to the choice of principles of triage in conditions of dire scarcity is the extent to which those principles provide grounds for, and perhaps encourage the cultivation of, hope that justice remains a possibility for generations subsequent to those for whom it is not possible, in virtue of the current generation's failure to curb its emissions.

Conclusion

The subject matter of this chapter has been bleak. And it may – depressingly – turn out to be the chapter in the book with most relevance to people fifty years hence, once talk of precautions has melted away, and the climate change horse has bolted out of the intergenerational-compensation-fund door. Just one indication of why this should be so is that, on current trends since 2000, global emissions are increasing at a rate of 3 per cent per year, whereas the most pessimistic IPCC scenario projects emissions growth rates of 3.4 per cent at most. Furthermore, some scenarios modelling recent energy-use trends project annual global emissions to be almost double current volumes, that is, 11 per cent higher than the worst-case business-as-usual IPCC scenarios.[38] There is, it seems, no good news about climate change. In the next chapter, and by way of conclusion, I offer an account of one way to motivate people to do intergenerational climate change justice that draws on a Rawlsian liberal conception of what unites generations in the pursuit of justice, with the proviso that these reflections may well be too little, and far too late.

6

CONCLUSION

Getting motivated in the last-chance saloon

INTERVIEWER: Do you think you'll ever give up flying?
SIENNA MILLER: I don't think I can at the moment (laughs). I'd love to … with the amount I've been doing it would be lovely to not to have to do quite so much but, you know, I love my job, it's wonderful to have a creative outlet, and I feel very lucky to be able to make a living out of what I love doing, but you know I can't avoid flying … start building boats (laughs) … rowing myself around … No, I can't see that in the foreseeable future. But I can start taking less baths.

(Interview with the actress Sienna Miller, spokesperson for Global Cool, a charity whose aim is to get each person to commit to reducing their CO_2 emissions by 1 tonne per year).[1]

If there are three facts about dealing with climate change over which no serious scientists, policy makers, or theorists (ought to) disagree, they are these:

(1) The nature of the climate change problem means that action must be taken now. We have a shrinking window of opportunity (a decade, perhaps two, at best) to make changes that will limit the temperature rises that our forbears' increased greenhouse gas emissions have already made inevitable.[2]

(2) Failure to take this action now will likely doom to failure action taken by future generations to limit global temperature rises to acceptable levels, for two reasons. First, as I have repeatedly emphasised throughout this book, we know that as greenhouse gas emissions rise climate change is likely to accelerate in virtue of a variety of positive feedbacks which increase warming; for example, as the Amazon rainforest dies off and its vegetation rots it releases CO_2 and thus the forest is transformed from being one of the Earth's major CO_2 sinks to a major CO_2 source, exacerbating the problem of spiralling greenhouse gas emissions and causing further warming. Second, as warming accelerates the likelihood increases

that the conditions we will bequeath to future generations through our failure to control emissions will be so hard for them – possibly catastrophic – that coordinated, concerted political action to achieve solutions (let alone *just* solutions) becomes less and less feasible. This is so because the size of the climate change problem will be greatly increased for them, while their resources will be greatly diminished. As discussed in the previous chapter, this is a nasty double whammy which makes it likely that the best that can be achieved by people in these circumstances are principles of triage, and hope for a future return to justice.

(3) Those not yet born cannot be held responsible for whether the current generation takes, or fails to take, this action; are powerless to affect this decision; and are vulnerable to its effects. In virtue of its time scale, the climate change problem is essentially a problem of intergenerational justice. Those who will be most harmed by a failure to lower emissions will be those who had no part in the choice not to act to lower emissions.

What (1), (2), and (3) mean is that the current generation are uniquely placed in human history: the choices we make now – in the next 10–20 years – will alter the destiny of our species (let alone every other species) unalterably, and forever. Generations prior to us did not know the havoc they were wreaking, and it will be too late for those who come after us. The current generation is drinking in the last-chance saloon of climate change mitigation, and closing time is near.

There is one further, crucial fact about greenhouse gas reductions which adds the final layer of intractability to the climate change problem: the changes required to make the required reductions are root and branch, and swingeing. These requirements bear most heavily on the people who have the most to lose in quantitative terms, i.e. inhabitants of developed, industrialised nations who enjoy all the comforts of the modern world, and who have the most resources available for short-term adaptation to global warming.[3] And by extension, their politicians have least incentive to enact measures to achieve these greenhouse gas cuts. This means that the most important question of climate change politics at the present time – the question that dwarfs every technical debate about the details of policy, philosophical dispute about the source of value, and green tirade about the evils of Western consumerism – is this: how can we motivate luxury emitters (the rich) to accept, indeed to initiate, the changes to their lifestyles that are required for bottom-line greenhouse gas emissions reductions? I shall refer to this as 'the motivation problem'.[4]

We might think that merely pointing out the harm that would be done to future generations by a failure to reduce emissions should be sufficient to motivate the current generation, including its subset of richest, to make these reductions. Indeed, it should: imposing these avoidable costs on the unborn is a paradigm case of injustice insofar as it constitutes exploitation of a powerless and vulnerable group.[5] However, merely pointing this out to the current generation – especially its wealthiest members – is unlikely to have sufficient motivating force to generate the required changes.[6] There is little evidence that merely raising awareness of environmental problems is sufficient to change individual behaviour in the long term,[7] although it may have short-term effects.[8]

At this point let me recap on points made in Chapter 2 in connection with the motivation problem. In *A Theory of Justice* Rawls argues that parties in the original position would be motivated to adopt a principle of savings in order to deliver justice to future generations as a result of reflection on what they would feel entitled to claim from their genetic or familial predecessors.[9] Rawls later – and rightly – abandoned this approach in favour of the claim that '[t]he [just savings] principle … is one the members of any generation (and so all generations) would adopt as the principle they would want preceding generations to have followed, no matter how far back in time'.[10] In Chapter 2 I presented a vision of a 'social union of social unions' united in the pursuit of justice with intergenerational scope that resonates with the reasoning of parties in the original position about the just savings principle as recast by the later Rawls. Part of the reason why parties would adopt a principle they would want all previous generations to have adopted is that they are part of a transhistorical union in a position to enjoy the good of social cooperation to achieve justice. Knowing this, parties in the original position will choose a principle of intergenerational justice that enables them – whatever their generation – to enjoy this good. Knowing that they are members of some generation located in time, although without knowing what generation this is, they will choose a principle of intergenerational justice that delivers this social good wherever in time they are located. The just savings principle is justified in this way because it aims at the creation, and then the maintenance, of social and economic conditions making justice possible. As Rawls puts it,

> [E]stablishing and successfully conducting reasonably just (though of course always imperfect) democratic institutions over a long period of time, perhaps gradually reforming them over generations, though not, to be sure, without lapses is a great social good and appreciated as such. This is shown by the fact that people refer to it as one of the significant achievements of their history.[11]

The account of intergenerational relations as cooperative delivers justifying reasons to support principles of intergenerational justice. However, reasons which justify but are not fit to motivate make for an impoverished political justification. Successful justifying reasons must be fit to motivate people to act for these reasons; they must be action guiding. Justifying reasons must be motivationally adequate because political justification is a practical activity: it ultimately aims to convince people that either they ought to restrain themselves according to its principles, or the coercion used to restrain those who do not restrain themselves is legitimate. A reason which was not fit to motivate people to act could not be a practical reason. The construction of political justifications must have an eye both on the question of whether people *could* act for the reasons it offers, and on the question of what it would take for these reasons to become reasons upon which people *do* act.[12]

The account of intergenerational relations as cooperative delivers justifying reasons for principles of intergenerational justice upon which people could act, and elaboration of why this is so reveals a Rawlsian moral psychology that tells a plausible story about how these reasons could become reasons upon which people do act. In brief, the argument will be that cooperative relations of intergenerational justice in general serve as a social basis of self-respect in two ways:

(a) for the person who cooperates;
(b) for future persons benefited by this cooperation insofar as it is a condition of their being in a state of progress with respect to pursuit of the social good of intergenerational justice.

Failure by a person acting in pursuit of their self-interest to do their part in securing these bases provides the following motivating reasons:

(a) shame at her failure to make some contribution to the achievement of intergenerational justice;
(b) a desire not to make future generations ashamed of her.

I shall argue that these general considerations about intergenerational justice and motivating reasons connected with self-respect, shame, and a commitment to act justly have special salience in the last-chance saloon circumstances of climate change in which the current generation is situated.

Let me begin by elaborating further the comments made in Chapter 2 about the role of self-respect and its social bases in Rawls' thought. Rawls claims that self-respect is 'perhaps the most important primary good';[13] self-respect combines self-value with self-confidence in a way essential for well-being and practical reasoning aimed at the good life, understood as plural in content but limited by non-subservience, and by congruence between self-conception and self-expression.[14] Many life plans informed by diverse conceptions of the good can provide a context for self-respect so long as those who pursue them realise significant self-set goals, and consider themselves to be subservient to no one else. The significance of self-respect as a *natural* primary good is registered in the inclusion of the *social* bases of self-respect on the list of social primary goods, which provide the currency of justice in Rawls' theory. The social bases of self-respect are a set of opportunities for people (with a diversity of reasonable conceptions of the good informing their life plans) to develop self-respect through their behaviour and interaction with others.

> [S]elf-respect depends upon and is encouraged by certain public features of basic social institutions, how they work together and how people who accept these arrangements are expected to (and normally do) regard and treat one another. These features of basic institutions and publicly expected

> (and normally honoured) ways of conduct are the social bases of self-respect.[15]

Although self-respect cannot be provided directly with political measures,[16] political and social conditions can present severe obstacles to it such that whole swathes of persons are systematically damaged. Caste systems, segregation, class divisions, rigid gender roles, and other forms of social hierarchies are the most obvious forms of social organisation that cause such harm. In these cases people in less privileged groups are (*inter alia*) viewed and treated as having less value than those who oppress and dominate them; and this can affect the self-conceptions of the oppressed in ways damaging to self-respect (evinced in the 'Uncle Tom' figure).[17] In addition, more subtle forms of injustice related, in particular, to distributional inequalities can also deprive people of the social bases of self-respect through the social exclusion, and the inequalities of opportunity to develop and refine cherished conceptions of the good, that they cause.

Following Joshua Cohen, we can classify the social bases of self-respect as: 'resource' bases, which are practical means enabling people to develop themselves according to their non-subservient self-conceptions; and 'recognitional' bases, constituted by the attitudes of others towards the person that support and foster her self-respect. Both resource and recognitional bases of self-respect can be found either in the 'associational' or in the 'framework' conditions of society. The associational bases of self-respect are communities of shared interests in civil society providing conditions of reciprocal esteem, wherein our achievements are recognised by others whose talents we value.[18] The framework bases inhere in the basic structure of society – its central political and social institutions – and are constituted by 'the framework of institutions and associated forms of public argument [which] support and foster the associational conditions'.[19] The other social primary goods (basic liberties, freedom of movement and free choice of occupation, powers and prerogatives of offices and positions of responsibility in the basic structure of society, and income and wealth), and their patterns of distribution, all serve as social bases of self-respect either insofar as they provide tangible resources in civil society or the basic structure for the pursuit of self-respect related activities, or insofar as they enable self-respect supporting (less tangible) mutual recognition between persons qua citizens or qua 'private' members of various communities of shared interests in which they participate, or both. For example, when guaranteed by the basic structure of society, equal freedom of association is a resource basis of self-respect because it enables persons to participate in any community of shared interest they see fit. It is also a recognitional basis because this guarantee symbolises citizens' recognition of the validity of one another's pursuit of conceptions of the good despite their deep disagreement with respect to the content of these conceptions, and because it gives persons qua private individuals access to the reciprocal esteem that can be found in the communities of shared interests in which they pursue their conceptions of the good.[20]

Of most relevance to the motivation problem are the recognitional bases of self-respect as found in the framework conditions (basic structure) of society. A person who finds her self-respect in associational membership and the activities that go with it – in, as Rawls sometimes calls such communities, social unions – normally also finds her self-respect supported when those with whom she is not closely associated as a private person, but with whom she shares citizenship, include her in public justificatory debate about justice and the terms of social cooperation by offering her justificatory reasons in public reason, and by taking seriously the reasons she offers and sincerely takes to be presentable in public reason. This form of inclusion can serve as a further recognitional basis of self-respect. The reason why mutual recognition between citizens qua citizens can constitute a social basis of self-respect is that self-respect depends on a person's conception of herself as worthy of being given justifying reasons by any other person capable of having expectations of her that she act in a certain way in social cooperation (this is how the non-subservience essential to self-respect is manifested in a person's identity qua citizen). The fact that a person is in a political community means that her fellow citizens will have expectations of her that she act in accordance with principles they decide upon in justificatory debate, i.e. just principles of social cooperation. In other words, social union can consist in the private activities of individuals united in the pursuit of specific goals by a shared conception of the good, or it can consist in the joint pursuit of justice by citizens. As Rawls makes explicit in his final work: '[a] … reason political society is a good for citizens is that it secures for them the good of justice and the social bases of mutual- and self-respect'.[21] And (less explicitly) in his first work:

> [A] well ordered society (corresponding to justice as fairness) is itself a form of social union. Indeed, it is a social union of social unions. Both characteristic features [of a social union] are present: the successful carrying out of just institutions is the shared final end of all the members of society, and these institutional forms are prized as good in themselves.[22]

When a person is excluded from justificatory debate about justice either by having her reasons ignored or by not being offered reasons by others at all, her status as a person worthy of giving and receiving justificatory reasons is undermined, and the recognitional bases of her self-respect as found in the basic structure of society come under threat. Furthermore, those who so deplete the bases of self-respect for others *ipso facto also* damage their own bases of self-respect by destroying the possibility of a social union of social unions with others united in the pursuit of justice. Of course, in both cases – those whose bases are damaged by others, and those who damage their own bases – the harmed parties may be unaware of the harm they suffer, and may enjoy healthy levels of self-esteem. But the warm glow of self-esteem is neither necessary nor sufficient for self-respect,[23] and provision of its social bases is not the business of principles of justice.[24]

The final – and crucial, for my purposes – part of the puzzle is to note that the social union of social unions providing recognitional bases of self-respect for citizens in the basic structure is *intergenerational*.

> [A] just constitutional order, when adjoined to the smaller social unions of everyday life, provides a framework for these many associations and sets up the most complex and diverse activity of all. In a well-ordered society each person understands the first principles that govern the whole scheme as it is to be carried out over many generations; and all have a settled intention to adhere to these principles in their plan of life … Everyone's more private life is so to speak a plan within a plan, this superordinate plan being realised in the public institutions of society.[25]

The intergenerational cooperative pursuit of justice is central to an intergenerational social union of social unions providing a recognitional basis of self-respect in the basic structure of the society of each generation that participates in it. If self-respect is the most important primary good – the one without which '[a]ll desire and activity becomes empty and vain, and we sink into apathy and cynicism'[26] – then delivering justice to future generations is in our interests as well as theirs, for without it we make intergenerational union with them impossible. If – as I have argued throughout this book – a commitment to intergenerational justice requires that we take radical action to reduce our emissions, and to provide funds to compensate future generations for harms we may be making inevitable for them, then the conditions of our own self-respect provide motivating reasons for us to deliver climate change justice to future generations.

Two questions remain, which can be stated as objections.

(1) Even if this account provides a powerful justifying reason for doing intergenerational justice with respect to climate change, the motivating force of the reasons it identifies still remains unclear.
(2) There is no reason to think that this account gives especially powerful motivating reasons for doing intergenerational climate change justice over other forms of intergenerational justice.

Turning first to the objection in (1), the motivating force of reasons for action related to the preservation of the social bases of self-respect is best revealed by consideration of self-respect's opposite, shame. Rawls states that 'shame is the emotion evoked by shocks to our self-respect',[27] and distinguishes 'natural' from 'moral' shame in terms of failures to exhibit different types of 'excellences' (qualities prized by people in social union with one another) related to self-respect. Natural shame is experienced in virtue of failings beyond our control which nevertheless matter to us from the perspective of the values, etc. around which self-respect is built;[28] for example, clumsiness, stammering, or obesity. Arguably, natural shame is a corrosive emotion with roots in social practices that aim at stigmatising some in order to normalise – and thus empower – others, and as such it should have no role in any liberal politics of mutual respect.[29] In any case, this form of shame is not apposite to my argument here. Instead, I shall appeal to Rawls' 'moral' shame: 'someone is liable to moral shame when he prizes as excellences of his person those virtues that his plan of life requires and is framed to encourage. He regards the virtues ... as properties that his associates want in him and that he wants in himself.'[30]

The experience of moral shame is generally more painful than that of natural shame, and is for that reason more motivating. Experiencing natural shame, we can comfort ourselves with the thought that we are not to blame for the defect that causes it; and sometimes we can thereby overcome natural shame altogether. But no such comfort is available when we experience moral shame (and nor ought it to be, in many cases):

> 'we are struck by the loss to our self-esteem and our inability to carry out our aims: we sense the diminishment of self from our anxiety about the lesser respect that others may have for us and for our disappointment with ourselves for failing to live up to our ideals'.[31]

Moral shame provides powerful motivating reasons for animals as social as we are. The strongest of us will, most of the time, do as we ought in order to avoid it, and the concomitant damage it does to our standing in the social unions we treasure as social bases of self-respect. The weaker of us will go to great lengths to hide our failures from fellow members of these unions, and as a result will experience constant anxiety at the possibility of being exposed, in the knowledge that the subterfuge will provide yet further reasons for moral shame. A tiny number of us simply lack the capacity for moral shame and, when discovered, are rightly treated with extreme caution, and are offered treatment, or at least partial quarantine. Most of us lie somewhere between the weak and the strong on this spectrum.

Given how cooperative union in the pursuit of intergenerational justice serves as a social basis of self-respect for persons in any time slice, a failure to do justice to future generations provides reasons for moral shame that have the potential to be powerfully motivating. Such failures also provide justifying reasons for moral guilt insofar as we have failed to do the right thing but, as Rawls notes, the perspective in this case is different. Experiencing guilt, 'we focus on the infringement of the just claims of others and the injury we have done to them, and on their probable resentment or indignation should they discover our deed',[32] whereas shame looks to the damage done to the bases of our own self-respect in virtue of how our failure will alter the attitudes of others towards us. Guilt looks outwards; moral shame, inwards. It is easy for a person to overlook or ignore the likely resentment or indignation of others whose just claims she overrides, and to whom she does injury – i.e. to ignore the grounds she has for guilt – especially if those she harms are poor and powerless and she has a lot to gain from doing them harm. It is far less easy to stand apart and preserve self-respect intact in the face of the judgements of others that one is unworthy of respect, and should be shunned for that reason. Despite tales of honour among thieves etc., upon achieving sufficient affluence most wrongdoers seek assimilation and acceptance by those who at least appear to be just and legitimate, and not only because this makes it easier for them to continue to do wrong.[33] A large part of my argument in this book has been that we have grounds for guilt with respect to our greenhouse gas emissions insofar as they will impact on future generations. But with respect to the motivation problem, it is moral shame that holds more promise, and which is an appropriate response to exactly the same emissions.

To those who think that reflections about self-respect and shame at this level of abstraction have no bearing on getting people motivated in the real world, consider the results of a 2005 study by the US Advertising Council and the Environmental Defense Fund examining the impact of different messages aiming to get people to take action on climate change.

> Participants responded best to [the message] which suggested that taking action would be 'doing good', which appealed to their innate goodness. The second best response was to a message that appealed to logic and responsibility. Positive psychologists confirm that people have a deep desire to live a 'good' or 'meaningful' life, in which they derive gratification from exhibiting their strengths, talents, and virtues, and use these skills and strengths to belong to and serve a larger purpose.[34]

When we fail to act according to principles of justice within any time slice we risk destroying the associational bases of self-respect provided by our co-citizens. But because the social union of social unions providing these bases is inter- as well as intragenerational, we also damage these bases as provided in the attitudes of future people – whom we injure through our failures – towards us; just as persons need not be physically present to us for the bases of self-respect they make possible to be destroyed by us, neither need they be temporally present.

Knowing that posterity will fail to respect us, will judge us as unjust, indeed, will be ashamed of being descended from us, gives powerful motivating reasons for doing intergenerational justice. But the failure to do intergenerational justice with respect to climate change caused by the motivation problem described at the start of this chapter has features that make it acute as a source of reasons for moral shame. Highlighting these features gives additional force to the thought that we have powerful motivating reasons to reduce our greenhouse gas emissions to levels that will make climate change manageable for future generations. And this is particularly true of luxury emitters, who can make the greatest contributions to this intergenerational project, and who – in virtue of having most to lose – are often the least motivated among us.

Recall that the current generation is uniquely placed in human history: what we do in the next 10–20 years will change circumstances for future generations either so as to make climate change a problematical, unfortunate, but ultimately tractable fact of life, or so as to make it the biggest disaster ever to have hit the species. Those who make the most malleable contributions to global greenhouse gas emissions are luxury emitting individuals (and countries and corporations): the 'Sienna Millers' of the world. The motivation problem is most forceful with respect to this group because their wealth renders them most able to adapt (at least in the short term) to the conditions that climate change could create. There are three facts about this group worth noting. First, they are in a position to make drastic cuts to their greenhouse gas emissions without reducing themselves to the level of the current and future poor – the subsistence emitters – who will be hardest hit by climate change: making these cuts is consistent with their remaining at the top of the global heap in terms of advantage. Second, these cuts could be effected immediately by relatively straightforward changes to lifestyle that are easily made and impose little cost, e.g. ceasing to travel by air. Third, the reason for not making these changes is, at best, lethargy and wilful myopia, or at worst, self-interested exploitation of future generations and the current poor. Future generations who will suffer if the current generation – especially the rich of the current generation – fail to reduce their greenhouse gas emissions will know these facts, and will with good reason be ashamed to be descended from such people. If we fail to make the cuts required within the next couple of decades it is no exaggeration to say that we, the particular cohort of people alive now, will have failed to do justice to the whole human race. For shame.

NOTES

1 Introduction

1 Solomon, S., D. Qin, M. Manning, Z. Chen, M. Marquis, K.B. Averyt, M. Tignor and H.L. Miller (eds), *Climate Change 2007: The Physical Science Basis. Contribution of Working Group I to the Fourth Assessment Report of the Intergovernmental Panel on Climate Change*, Cambridge: Cambridge University Press, 2007. Online. HTTP: <http://www.ipcc.ch/publications_and_data/ar4/wg1/en/contents.html> (accessed 10 August 2010), pp. 96–103 and pp. 465–83. See also <http://www.ipcc.ch/publications_and_data/ar4/wg1/en/faq-1-1.html> and http://www.ipcc.ch/publications_and_data/ar4/wg1/en/faq-6-2.html (accessed 30 July 2010).

2 See ibid. p. 465.

3 There are, of course, those who deny that current climate change has man-made causes, but they are a shrinking minority, and their claims have no credence in the face of the IPCC Assessments. For denunciation of this, and the many other claims of climate deniers, see C. Beck, 'How To Talk to a Climate Sceptic', *A Few Things Ill Considered*, 2008. Online. HTTP: <http://scienceblogs.com/illconsidered/2008/07/how_to_ talk_to_a_sceptic.php> (accessed 10 August 2010).

4 The thermohaline circulation is the global network of currents driven by the varying salinity of sea water throughout the oceans, temperature at the surface, injections of freshwater, wind, and other factors.

5 See S. Solomon et al., *Climate Change 2007: The Physical Science Basis*, pp. 114–15.

6 See R.K. Pachauri and A. Reisinger (eds), *Climate Change 2007: Synthesis Report. Contribution of Working Groups I, II and III to the Fourth Assessment Report of the Intergovernmental Panel on Climate Change*, Geneva: IPCC, 2007. Online. HTTP: <http://www.ipcc.ch/publications_and_data/ar4/syr/en/contents.html> (accessed 10 August 2010), p. 36.

7 See ibid., p. 36.

8 See ibid., p. 37.

9 See J. Houghton, *Global Warming: The Complete Briefing*, 4th edn, Cambridge: Cambridge University Press, p. 76.

10 All data taken from ibid., pp. 76–77.

11 The IPCC was established by the United Nations (UN) and World Meteorological Organization in 1988, and reports on the causes, impacts, and options for tackling climate change every five years or so. It does not itself conduct research, but rather

reviews and synthesises the research work of scientists across a range of fields related to climate change. This work is done by thousands of scientists across the globe on a voluntary basis.

12 Pachauri and Reisinger (eds), *Climate Change 2007: Synthesis Report*, p. 39.

13 Ibid., p. 40.

14 See ibid., Fig. 2.1, p. 36.

15 Some prominent scientists are starting to question the 2°C threshold. For example, Anderson and Bows argue that the 2°C target is 'a dangerously misleading basis for informing the adaptation agenda' because achieving the emissions reductions necessary for stabilisation at 2°C is so unlikely. See K. Anderson and A. Bows, 'Reframing the Climate Change Challenge in the Light of Post-2000 Emission Trends', *Philosophical Transactions of the Royal Society A* 366, 2008, pp. 3863–82. See also J. Hansen, 'Target Atmospheric CO2: Where Should Humanity Aim?', *Open Journal on Atmospheric Sciences* 2, 2008, pp. 217–31. Hansen argues for a CO_2 stabilisation target of 350ppm, which is below current levels.

16 M. Lynas, *Six Degrees: Our Future on a Hotter Planet*, London: HarperCollins, 2007, pp. 270–71.

17 See Pachauri and Reisinger (eds), *Climate Change 2007: Synthesis Report*, Table 5.1, p. 67. The percentage reduction is relative to emissions in 2000, and the temperature increase is relative to pre-industrial levels. Note that this increase is calculated using the IPCC's best estimate of climate sensitivity of 3°C. For discussion see note 18.

18 'Climate sensitivity' is 'the equilibrium change in the annual mean global surface temperature following a doubling of the atmospheric equivalent carbon dioxide concentration' (Pachauri and Reisinger (eds), *Climate Change 2007: Synthesis Report*, Appendix, p. 78). Very crudely, the lower the value given for climate sensitivity, the less responsive to anthropogenic greenhouse gas emissions the climate is modelled as being. The IPCC'S Fourth Assessment Report states that

> 'climate sensitivity is likely to be in the range of 2 to 4.5°C with a best estimate of about 3°C, and is very unlikely to be less than 1.5°C. Values substantially higher than 4.5°C cannot be excluded, but agreement of models with observations is not as good for those values'

(ibid., p. 38). If climate sensitivity is at the top of the range specified by the IPCC, then global average temperature is likely to rise higher than the 2°C–2.4°C indicated in the IPCC's best-case scenarios wherein emissions peak early and decline rapidly.

19 Ibid., pp. 72–73.

20 The IPCC describes this positive feedback as follows:

> 'Future *climate change* induced by atmospheric emissions of *greenhouse gases* will impact on the global *carbon cycle*. Changes in the global carbon cycle in turn will influence the fraction of anthropogenic greenhouse gases that remains in the atmosphere, and hence the atmospheric concentrations of greenhouse gases, resulting in further climate change.'

Ibid., Appendix, p. 78.

21 See ibid., p. 67. See also Houghton, *Global Warming*, pp. 314–15, for typically restrained, but potent, words of caution with respect to the 2°C threshold.

22 The IPCC stabilisation scenarios are informed by projections of the stabilisation levels and range of temperature increases at equilibrium found in families of scenarios (the so-called 'SRES' scenarios used by the IPCC to model possible future climate change (see N. Nakicenovic and R. Swart, *Special Report on Emissions Scenarios*,

Cambridge: Cambridge University Press, 2000. Online. HTTP: <http://www.ipcc.ch/ipccreports/sres/emission/index.php?idp=0> (accessed 10 August 2010). Each family of SRES scenarios – A1, A2, B1 and B2 – makes different sets of assumptions about future demographic, economic, technological, social and environmental developments, woven together with a narrative storyline. The SRES scenarios exclude the sorts of climate catastrophes (and their associated positive feedbacks) identified by Lynas, and the IPCC does not assign a probability to any of them. See ibid., p. 3. Note that scenarios have been developed since this report that have a far higher upper limit for emissions of $GtCO_2$/yr than even the worst-case stabilisation category VI. The AR4 does not discuss the post-SRES scenarios.

23 'Stabilisation' is the point at which concentration of CO_2-eq – or any greenhouse gas – emissions reaches a constant (see Pachauri and Reisinger (eds), *Climate Change 2007: Synthesis Report*, Appendix, p. 87). 'Equilibrium' is the point at which global average temperature increases level off, and cease to rise. As the IPCC notes, 'global average temperature at equilibrium is different from expected global average temperature at the time of stabilisation of greenhouse gas concentrations due to the inertia of the climate system' (ibid., p. 67, note (e) to Table 5.1). For the majority of IPCC scenarios stabilisation is achieved between 2100 and 2150; the inertia of the climate system means equilibrium is achieved some time later, 'over a few centuries', ibid., p. 67. n. 30.

24 The IPCC sometimes gives temperature increases at equilibrium relative to the pre-industrial period, and sometimes relative to the period 1980–99 – the difference between the two is 0.5°C, and I shall indicate which figure is being used, where applicable.

25 See ibid., p. 66. Note also the following statement made in *The Stern Review*:

> 'Under a Business-as-Usual scenario, the stock of greenhouse gases could more than treble by the end of the century, giving at least a 50% risk of exceeding 5°C global average temperature change during the following decades. This would take humans into unknown territory. An illustration of the scale of such an increase is that we are now only around 5°C warmer than in the last ice age'.

(*The Stern Review*, Executive Summary, p. iv). And see also Anderson and Bows, 'Reframing the Climate Change Challenge in the Light of Post-2000 Emission Trends' for analysis suggesting that stabilisation at anything under 650ppm CO_2-eq (correlated with a temperature increase of at least 4°C) is improbable.

26 Four of the six stabilisation categories used in the IPCC'S Fourth Assessment Report have stabilisation levels for CO_2-eq above 535ppm (correlating with a minimum temperature increase at equilibrium of 2.8°C, using the IPCC's best estimate of climate sensitivity).

27 See Pachauri and Reisinger (eds), *Climate Change 2007: Synthesis Report*, p. 45.

28 See V. Pope, 'The Scientific Evidence for Early Action on Climate Change', Met Office. Online. HTTP: <http://www.metoffice.gov.uk/climatechange/policymakers/action/evidence.html> (accessed 10 August 2010).

29 See Met Office, 'Avoiding Dangerous Climate Change', Met Office: 2008. Online. HTTP: <http://www.metoffice.gov.uk/publications/brochures/cop14.pdf> (accessed 10 August 2010). The Met Office graphic at the following location makes the scenarios vivid. Online. HTTP: <http://www.metoffice.gov.uk/climatechange/policymakers/action/emissions_270908.pdf> (accessed 19 August 2010).

30 See ibid., p. 13. The Met Office 'business-as-usual' scenario utilises IPCC assumptions informing its 'A2' storyline and scenario family, e.g. that global population grows, and that rates of economic growth and technological change are relatively slow. The Met Office scenario predicts temperature rises 1°C higher than the IPCC A2 scenario results because it assumes that the emission of aerosols – which have a cooling effect on the atmosphere – will decline.

31 See Pachauri and Reisinger (eds), *Climate Change 2007: Synthesis Report*, p. 51.

32 See S. Manabe, R.T. Wetherald, P.C.D. Milly, T.L. Delworth, R.J. Stouffer, 'Century-scale Change in Water Availability: CO2 Quadrupling Experiment', *Climatic Change* 64, 2004, 59–76.

33 Lynas, *Six Degrees*, pp. 208–9.

34 'The linear warming trend over the 50 years from 1956 to 2005 (0.13 [0.10 to 0.16]°C per decade) is nearly twice that for the 100 years from 1906 to 2005'. Pachauri and Reisinger (eds), *Climate Change 2007: Synthesis Report*, p. 30.

35 Ibid., pp. 30–33.

36 Throughout this book I limit myself to considerations of how climate change could impact on human beings, and I use an anthropocentric theory of value to assess these impacts and the justification of various principled approaches to prevention and adaptation. However, it is obviously the case that climate change will negatively impact on nonhuman animals and other species, and it is not obvious that an anthropocentric theory of value is well placed to assess these impacts (or, for that matter, impacts on human beings). Consideration of the harm climate change will cause to animals and other species, and defence of the anthropocentric approach used here, is beyond the scope of the present enquiry. For discussion see T. Hayward, 'Anthropocentrism: A Misunderstood Problem', *Environmental Values* 6, 1997, 49–63. For a powerful defence of anthropocentrism see O. O'Neill, 'Necessary Anthropocentrism and Contingent Speciesism', *Proceedings of the Aristotelian Society* Supplementary Vol. 72, 1998, 211–28.

37 M.L. Parry, O.F. Canziani, J.P. Palutikof, P.J. van der Linden and C.E. Hanson (eds), *Contribution of Working Group II to the Fourth Assessment Report of the Intergovernmental Panel on Climate Change*, Cambridge: Cambridge University Press, 2007. Online. http://www.ipcc.ch/publications_and_data/publications_ipcc_fourth_assessment_report_wg2_report_impacts_adaptation_and_ vulnerability.htm (accessed 10 August 2010), p. 13.

38 Ibid., p. 13.

39 Ibid., pp. 13–14.

40 Ibid., p. 14.

41 According to one study, there is a causal relationship between heat and violence. The study suggests that any increase in average global temperature is likely to be accompanied by an increase in violent aggression. Indeed, it suggests that current models predict a rise of about 24,000 assaults or murders in the US every year for every increase of 2 degrees Fahrenheit in the average temperature. See C.A. Anderson, 'Heat and Violence', *Current Directions in Psychological Science* 10(1), 2001, 33–38.

42 M.L. Parry et al. (eds), *Contribution of Working Group II to the Fourth Assessment Report of the Intergovernmental Panel on Climate Change*, pp. 14–15.

43 Ibid., p. 15.

44 Ibid., p. 15.

45 See 'Millions Say It Is Too Much Effort To Adopt Greener Lifestyle', *The Guardian*, 15 August 2007.

46 See Pew Research Center, *Luxury or Necessity? Things We Can't Live without: The List Has Grown in the Past Decade*, Pew Research Center: A Social Trends Report, 2006.

47 A classic discussion is P. Slovic, 'Perception of Risk', *Science*, New Series 236(4799), 1987, 280–85.

48 See X. Zhao and R.W. Belk, 'Politicizing Consumer Culture: Advertising's Appropriation of Political Ideology in China's Social Transition', *Journal of Consumer Research* 35, 2008, 231–44.

49 Pew Research Center, *Luxury or necessity?*

50 See <http://www.simpleliving.net/> (accessed 10 August 2010).

51 Literature reviews undertaken by the IPCC confirm this, and other, criticisms of the Kyoto Protocol. See B. Metz, O.R. Davidson, P.R. Bosch, R. Dave and L.A. Meyer (eds), *Climate Change 2007: Working Group III: Mitigation of Climate Change*, Cambridge: Cambridge University Press, 2007. Online. HTTP: <http://www.ipcc.ch/publications_and_data/ar4/wg3/en/contents.html> (accessed 10 August 2010), p. 768.
52 Available at http://www.youtube.com/watch?v=QLmP40gYH7c&amp:feature=player_embedded#! (accessed 5 August 2010).
53 A book which does this well with respect to US climate change policy is S. Vanderheiden, *Atmospheric Justice*, Oxford: Oxford University Press, 2008.

2 A Rawlsian approach to climate change justice

1 A third feature of the reasonable person is that she has a 'determinate conception of the good', according to which she sets herself ends, and which is connected with 'a view of our relation to the world – religious, philosophical, and moral – by reference to which the value and significance of our ends and attachments are understood'. J. Rawls, *Political Liberalism*, New York: Columbia University Press, 1993, pp. 19–20.
2 Ibid., p. 19.
3 Ibid., p. 243.
4 J. Rawls, 'The Idea of Public Reason Revisited' in S. Freeman (ed.), *John Rawls: Collected Papers*, Cambridge, MA: Harvard University Press, 1999, p. 594.
5 Ibid., p. 594.
6 J. Rawls, *Justice as Fairness: A Restatement*, Cambridge, MA: Harvard University Press, 2001, pp. 33–34.
7 See Rawls, *Political Liberalism*, p. 63.
8 All quoted material in a.–e. in ibid., pp. 56–57. In this work Rawls includes a sixth burden of judgement related to the 'limited social space' created by institutions in which values can be realised; however, this burden of judgement is relegated to a footnote in *Justice as Fairness*.
9 See Rawls, *Political Liberalism*, pp. 54–58, and Rawls, *Justice as Fairness*, pp. 35–36.
10 Rawls, *Justice as Fairness*, p. 90.
11 J. Rawls, 'Justice as Fairness: Political not Metaphysical' in Freeman (ed.), *John Rawls: Collected Papers*, p. 395.
12 The term 'epistemic abstinence' is coined by Joseph Raz. See J. Raz, 'Facing Diversity: The Case of Epistemic Abstinence', *Philosophy and Public Affairs* 19, 1990, 3–46.
13 Rawls, *Justice as Fairness*, pp. 89–90.
14 See J. Cohen, 'For a Democratic Society' in Freeman (ed.), *The Cambridge Companion to Rawls*, Cambridge: Cambridge University Press, 2003.
15 See, for example, M. Bookchin, *Social Ecology and Communalism*, Oakland, CA: AK Press, 2007.
16 An intelligent discussion can be found in T. Hayward, 'Anthropocentrism: A Misunderstood Problem', *Environmental Values* 6, 1997, 49–63.
17 J. Rawls, *A Theory of Justice*, rev. edn, Oxford: Oxford University Press, 1999, p. 386. Rawls continually stressed the importance of self-respect and its social bases throughout his work. See ibid., pp. 92, 107, 443, 543–45; *Political Liberalism*, pp. 106, 203, 318, 319; *Justice as Fairness*, pp. 59–60.
18 Rawls, *A Theory of Justice*, rev. edn. p. 386.
19 Ibid., 386.
20 For a full discussion of the connection between self-respect and practical reason see C. McKinnon, *Liberalism and the Defence of Political Constructivism*, Basingstoke: Palgrave Macmillan, 2002.
21 J. Rawls, *A Theory of Justice*, Oxford: Oxford University Press, 1971, p. 319.

22 This taxonomy of the social bases of self-respect is inspired by Joshua Cohen, 'Democratic Equality', *Ethics* 99, 1989, 727–51. For a full discussion see McKinnon, *Liberalism and the Defence of Political Constructivism*, chapter 5.
23 Rawls, *A Theory of Justice*, rev. edn, p. 462.
24 See N. Rosenblum, *Membership and Morals*, Princeton, NJ: Princeton University Press, 2000.
25 See McKinnon, *Liberalism and the Defence of Political Constructivism*, §3.3 for full discussion.
26 See also C. Larmore, 'Public Reason' in Freeman (ed.), *The Cambridge Companion to Rawls*, pp. 373–74.
27 See Rawls, *Political Liberalism*, p. 152.
28 Ibid., pp. 252–53.
29 Ibid., p. 214.
30 Rawls, *Political Liberalism*, p. 214; see also Rawls, *Justice as Fairness* p. 91.
31 Rawls, *Political Liberalism*, p. 215.
32 One commentator who notes this is Larmore, 'Public Reason', pp. 381f.
33 See Rawls, *Justice as Fairness*, Part V.
34 Ibid., pp.89–90. See also Rawls, *Political Liberalism*, p. 220.
35 See Rawls, *Justice as Fairness*, p. 90
36 An important forthcoming paper related to discussions in this section is E. Anderson, 'Democracy, Public Policy and Lay Assessments of Scientific Testimony', *Episteme* 8(2), 2011, 144–64. Unfortunately, I became aware of this paper too late to include discussion of it.
37 Ibid., p. 92. See also Rawls, *Political Liberalism*, p. 220ff.
38 Rawls, *Justice as Fairness*, p. 92. These differences relate in part to 'how the nature (the aim and point) of each association is understood and the conditions under which it pursues its ends', ibid., p. 93.
39 See Rawls, *Political Liberalism* pp. 220–22. Here I gloss over Rawls' 'proviso' that public reason-based guidelines of enquiry do not prevent us from 'introduc[ing] into political discussion at any time our comprehensive doctrine, religious or non-religious, provided that, in due course, we give properly public reasons to support the principles and policies our comprehensive doctrine is said to support' (Rawls, 'The Idea of Public Reason Revisited', p. 584). Because the proviso retains the demand that all reasons offered in political debate are *ultimately* presented in public reason, it does not affect my argument here against Rawls' devotee-scientist analogy.
40 See C.A. Miller and P.N. Edwards, 'Introduction: The Globalization of Climate Science and Climate Politics' in C.A. Miller and P.N. Edwards, *Changing the Atmosphere: Expert Knowledge and Environmental Governance*, Cambridge, MA: MIT Press, 2001.
41 For discussion of the latter case see C. McKinnon, *Toleration: A Critical Introduction*, London: Routledge, 2005, chapter 7.
42 Rawls, *Justice as Fairness*, pp. 89–90. See also Rawls, *Political Liberalism*, p. 220.
43 Thanks to Tank Waddington for discussion on this point.
44 See 'Earth Heats Up as the Smog Clears', *New Scientist*, 14 May 2005, p. 27. See also http://www.bbc.co.uk/sn/tvradio/programmes/horizon/dimming_prog_summary.shtml (accessed 23 July 2010).
45 Rawls, *Justice as Fairness*, p. 90. In fact, it is doubtful whether there are any conclusions of science that are not controversial; see H.N. Pollack, *Uncertain Science … Uncertain World*, Cambridge: Cambridge University Press, 2003.
46 Rawls, *Justice as Fairness*, p. 90.
47 Here I follow O. Godard, 'Integrating Scientific Expertise into Regulatory Decision-Making. Social Decision-Making under Conditions of Scientific Controversy Expertise and the Precautionary Principle', EUI Working Paper RSC No. 96/6, 1992. See also J. Krebs and A. Kacelnik, 'Risk: A Scientific View' in *Science, Policy and Risk*,

London: The Royal Society, 1997, 31–44; A.E. Dessler and E.A. Parson, *The Science and Politics of Global Climate Change*, Cambridge: Cambridge University Press, 2006, chapter 2.

48 See 'Sceptics Forced into Climate Climbdown', *New Scientist*, 20 August 2005, p. 10. See also R. Gelbspan, *The Heat is On: The High Stakes Battle over Earth's Threatened Climate*, Reading, MA: Addison-Wesley Publishing Co., Inc, 1997.

49 The most recent calculations of the Christy team estimate a warming trend, albeit smaller than estimates based on data collected from surface thermometers. See Dessler and Parson, *The Science and Politics of Global Climate Change*, pp. 62–64; J. Christy, R.W. Spencer, W.B. Norris, W.D Braswell and D.E. Parker, 'Error Estimates of Version 5.0 of MSU-AMSU Bulk Atmospheric Temperatures', *Journal of Atmospheric Oceanic Technology* 20, 2003, 613–29.

50 See < http://www.ipcc.ch/organization/organization.htm > (accessed 23 July 2010).

51 A further interesting question is whether scientists are *required* to indicate to policy makers that *The Permissibility of Expertise* has been activated; this would give scientists far greater social responsibility than we, or they, normally suppose that they have, and perhaps more than many of them would want (although on this question see the website of the US Union of Concerned Scientists, an organisation committed to the use of 'independent scientific research and citizen action to develop innovative, practical solutions and to secure responsible changes in government policy, corporate practices, and consumer choices'; http://www.ucsusa.org/ (accessed 23 July 2010).

52 See R.K. Pachauri and A. Reisinger (eds), *Climate Change 2007: Synthesis Report. Contribution of Working Groups I, II and III to the Fourth Assessment Report of the Intergovernmental Panel on Climate Change*, Geneva: IPCC, 2007. Online. HTTP: <http://www.ipcc.ch/publications_and_data/ar4/syr/en/contents.html> (accessed 10 August 2010), p. 2.

53 See ibid., p. 2.

54 See ibid., p. 2.

55 See Dessler and Parson, *The Science and Politics of Global Climate Change*, p. 65.

56 See ibid., p. 65.

57 See ibid., p. 65.

58 See ibid., p. 65.

59 See N. Oreskes, 'The Scientific Consensus on Climate Change', *Science* 306 (5702), 2004, pp. 1686–87.

60 As the IPCC puts it,

> 'climate forcing by changes in solar irradiance and volcanism have likely caused fluctuations in global and hemispheric mean temperatures. Qualitative comparisons suggest that natural forcings produce too little warming to fully explain the 20th century warming … The indication that the trend in net solar plus volcanic forcing has been negative in recent decades … makes it unlikely that natural forcing can explain the increased rate of global warming since the middle of the 20th century.'

J.T. Houghton, Y. Ding, D.J. Griggs, M.Noguer, P.J. van der Linden, X. Dai, K. Maskell and C.A. Johnson (eds), *Climate Change 2001: The Scientific Basis. Contribution of Working Group I to the Third Assessment Report of the Intergovernmental Panel on Climate Change*, Cambridge: Cambridge University Press, 2001. Online. HTTP: <http://www.grida.no/publications/other/ipcc_tar/> (accessed 11 August 2010), p. 709. See in particular Fig. 12–17 at this location.

61 An analogy with Holocaust denial may be helpful here: that Holocaust deniers claim to be doing history should not lead policy makers to read their materials in deciding upon whether to build a monument to the victims of the Holocaust. Academic historians reject holocaust deniers as members of their discipline, and policy makers

must take this seriously. For discussion see C. McKinnon, 'Should We Tolerate Holocaust Denial?', *Res Publica* 12(4), 2007, 9–28.

62 A funny and informative short film elaborating aspects of this claim is 'Toe-to-Toe with Monbiot Part 1', in which Franny Armstrong (director of the film *Age of Stupid*) interviews George Monbiot on the subject of climate change deniers. Online. HTTP: http://www.ageofstupid.net/stupid-show (accessed 23 July 2010).

63 The UK Office of Science and Technology offers guidance to scientists to address this problem in its 2002 joint (with Research Councils UK) report 'Dialogue with the Public: Practical guidelines'. Online. HTTP: http://www.rcuk.ac.uk/cmsweb/downloads/rcuk/scisoc/dialogue.pdf (accessed 23 July 2010).

64 The graph originates in M.E. Mann, R.S. Bradley and M.K. Hughes, 'Northern Hemisphere Temperatures during the Past Millennium: Inferences, Uncertainties, and Limitations', *Geophysical Research Letters* 26(6), 1999, 759–62. Online. HTTP: <http://www.ltrr.arizona.edu/webhome/aprilc/data/my%20stuff/MBH1999.pdf> (accessed 19 August 2010).

65 See, for example, Alaska governor Sarah Palin's claims in the *Washington Post* that the emails exposed a 'highly politicized scientific circle' and that the scientists 'deliberately destroyed records, manipulated data to "hide the decline" in global temperatures, and tried to silence their critics by preventing them from publishing in peer-reviewed journals', 'Sarah Palin on the Politicization of the Copenhagen Climate Conference', *The Washington Post*, 9 December 2009. Online. HTTP: http://www.washingtonpost.com/wp-dyn/content/article/2009/12/08/AR2009120803402.html (accessed 26 July 2010). Rallies during Palin's 2008 campaign to become the Republican Presidential candidate were marked by chants of 'Drill, baby, drill'. For a comprehensive refutation of Palin's claims, and others like them, see F. Pearce, 'How the "Climategate" Scandal is Bogus and Based on Climate Sceptics' Lies', *The Guardian*, 9 February 2010. Online. HTTP: <http://www.guardian.co.uk/environment/2010/feb/09/climategate-bogus-sceptics-lies (accessed 26 July 2010).

66 The most comprehensive of these investigations was Sir M. Russell, 'The Independent Climate Change E-Mails Review', 2010. Online. HTTP: http://www.guardian.co.uk/environment/2010/jul/07/findings-muir-russell-review (accessed 11 August 2010). Extensive commentary on the affair can be found at F. Pearce, 'Climate Wars: Guardian Special Investigation', 2010. Online. HTTP: http://www.guardian.co.uk/environment/series/climate-wars-hacked-emails (accessed 11 August 2010). Recent evidence from polls suggests that the negative effect of Climategate on public opinion may not be as bad as many expected. See B. Ward, 'Has the Impact of Climategate Worn Off?', *New Scientist*, 14 February 2011. Online. HTTP: http://www.newscientist.com/blogs/thesword/2011/02/has-the-impact-of-climategate.html (accessed 21 March 2011).

67 J. Dryzek, *The Politics of the Earth: Environmental Discourses*, Oxford: Oxford University Press, 1997, p. 74.

68 See T. O'Riordan, 'The Precautionary Principle and Civic Science' in T. O'Riordan, J. Cameron, and A. Jordan (eds), *Reinterpreting the Precautionary Principle*, London: Cameron May Ltd., 2001, p. 100.

69 See ibid., p. 100.

70 Through the Human Fertilisation and Embryology (Research Purposes) Regulations 2001, the UK government granted licenses for embryonic stem cell research. However, it acknowledges that '[f]uture developments might make further research on embryonic stem cell cells unnecessary' and accepts that

> '[a]t an appropriate time, perhaps towards the end of the decade, the Government should undertake a further review of scientific developments, particularly of the progress of adult stem cell research and therapies, and of the development of stem cell banks, with a view to determining whether research on human embryos is still necessary'.

Department of Health, *Government Response to the House of Lords Select Committee Report on Stem Cell Research*, 2002. Online. HTTP: http://www.dh.gov.uk/assetRoot/04/01/97/47/04019747.pdf (accessed 23 July 2010), p. 7 and p. 14 respectively.

71 See Godard, 'Social Decision-Making under Conditions of Scientific Controversy', p. 9.

72 See < http://ipa.org.au/people/bio.asp?peopleid=4 > (accessed 5 September 2005).

73 J. Marohasy, 'Environmental Fundamentalism', *Policy* 20(3), 2004. Online. HTTP: <http://www.cis.org.au/policy/spring04/spring04–07.pdf> (accessed 5 September 2005).

74 Ibid., p. 41.

75 Marohasy herself claimed that she did not take 'an interest in who funds IPA'. See <http://www.sourcewatch.org/index.php?title=Institute_of_Public_Affairs#Funding> (accessed 23 July 2010).

76 See P. Harris, 'Bush Covers Up Climate Research', *The Observer*, 21 September 2003. Online. HTTP: <http://www.guardian.co.uk/climatechange/story/0,12374,1046389,00.html> (accessed 23 July 2010). For more on the Bush administration's treatment of the climate change problem see D.A. Brown, *American Heat*, Lanham, MD: Rowman and Littlefield Publishers Inc., 2002.

77 See Krebs and Kacelnik, 'Risk: A Scientific View' p. 39. Erstwhile 'creation scientists' are well aware of the albatross that the label has become; hence the repackaging of their ideas with the term 'intelligent design'.

78 See Godard, 'Social Decision-Making under Conditions of Scientific Controversy', pp. 12–14.

79 Ibid., p. 13.

80 Dryzek, *The Politics of the Earth*, p. 81.

81 See http://www.futuregenalliance.org/, (accessed 23 July 2010).

82 The Asia-Pacific Partnership on Clean Development and Climate – which brings together Australia, Canada, China, India, Japan, the Republic of Korea, and the US in an agreement to share clean technologies – is committed to the development of this technology; see <http://www.asiapacificpartnership.org/english/default.aspx> (accessed 23 July 2010). See also 'Special Report: Clean Energy', *New Scientist*, 3 September 2005, pp. 30–43.

83 Uncertainty is present in science when outcomes but not their probabilities are known; ignorance is present when outcomes are not known. More controversially, indeterminacy is present when ignorance is a permanent or endemic feature. See M. McGarvin, 'Science, Precaution, Facts and Value' in O'Riordan, Cameron and Jordan (eds), *Reinterpreting the Precautionary Principle*, p. 42. See also S. Yearley, 'The Environmental Challenge to Science Studies' in S. Jasanoff, G.E. Markle, J.C. Petersen, and T. Pinch (eds), *Handbook of Science and Technology Studies*, Thousand Oaks: Sage, 1994.

84 See McGarvin, 'Science, Precaution, Facts and Value', p. 43. See also Pollack, *Uncertain Science … Uncertain World*, chapter 2.

85 Such scepticism is real. A 1995 MORI poll showed that 62 per cent of the sample had not much or no faith in scientists working for government, and 49 per cent had not much or no faith in those working for industry. See O'Riordan, 'The Precautionary Principle and Civic Science', pp. 100–101.

86 See Godard, 'Social Decision-Making under Conditions of Scientific Controversy', pp. 31–32.

87 Krebs and Kacelnik, 'Risk: A Scientific View', p. 40.

88 D. Bellamy, 'Global Warming? What a Load of Poppycock!', *Daily Mail*, 9 July 2004. Online. HTTP: http://www.junkscience.com/july04/Daily_Mail-Bellamy.htm (accessed 23 July 2010). See also G. Monbiot, 'Junk Science', *The Guardian*, 10 May 2005. Online. HTTP: http://www.monbiot.com/archives/2005/05/10/junk-science/ (accessed 23 July 2010).

89 Rawls, *Justice as Fairness*, p. 159.
90 Rawls, *A Theory of Justice*, rev. edn, p. 266 and p. 258.
91 And not, note, for the sake of simply enriching later generations. See ibid., p. 257.
92 I qualify this sentence with 'at least' because, given Rawls' belief that a just basic structure is a necessary condition for the realisation of any of the principles of justice as fairness insofar as this secures background justice (see Rawls, *Political Liberalism*, pp. 265–69), it is not clear why the just savings principle does not take priority over *all* the other principles of justice.
93 Rawls, *A Theory of Justice*, rev. edn, p. 252.
94 Ibid., p. 255.
95 Ibid., p. 255.
96 Rawls, *Justice as Fairness*, p. 43.
97 Rawls, *A Theory of Justice*, rev. edn, p. 252.
98 Ibid., p. 254.
99 F. Gaspart and A. Gosseries, 'Are Generational Savings Unjust?', *Politics, Philosophy and Economics* 6, 2007, 193–217, pp. 202–3.
100 Rawls, *A Theory of Justice*, rev. edn, p. 72.
101 Ibid., p. 255.
102 Gaspart and Gosseries, 'Are Generational Savings Unjust?', p. 204.
103 Rawls, *A Theory of Justice*, rev. edn, p. 288.
104 Gaspart and Gosseries identify a significant exception to the closed principle, which shows that intergenerational maximin may sometimes require a positive rate of savings even for generations in the 'steady state'. This is 'predictable exogenous disadvantage', i.e. knowledge that a future generation will suffer as a result of a future event, e.g. a natural disaster such as a meteor collision (see 'Are Generational Savings Unjust?', p. 209). Gaspart and Gosseries do not elaborate on why this stands as an exception, but presumably the reasons are connected with the fact that this disadvantage to the future generation is a result of brute bad luck.
105 Although see J. Wolff and A. de-Shalit, *Disadvantage*, Oxford: Oxford University Press, 2007, for discussion.
106 See Rawls, *A Theory of Justice*, rev. edn, p. 121.
107 Rawls, *Justice as Fairness*, p. 107.
108 See Rawls, *A Theory of Justice*, rev. edn, p. 131.
109 J. Rawls, *A Theory of Justice*, Oxford: Oxford University Press, 1971, p. 291.
110 Ibid., p. 291.
111 Rawls, *A Theory of Justice*, rev. edn, p. 254.
112 Ibid., p. 255.
113 Ibid., p. 255.
114 Ibid., pp. 255–56.
115 For a different critique – that focuses on how this modification creates problems for thinking about justice within the family – see J. English, 'Justice between Generations', *Philosophical Studies* 31, 1977, 91–104.
116 Rawls, *A Theory of Justice*, p. x
117 Gaspart and Gosseries, 'Are Generational Savings Unjust?', p. 198.
118 Rawls, *Political Liberalism*, p. 204. This passage is repeated, almost word for word, in Rawls, *Justice as Fairness*, p. 201.
119 Rawls, *Justice as Fairness*, p. 202.
120 See Rawls, *Political Liberalism*, pp. 202–4.
121 See Rawls, *Justice as Fairness*, p. 13.
122 See J. Thompson, *Intergenerational Justice: Rights and Responsibilities in an Intergenerational Polity*, New York: Routledge, 2009, especially pp. 39–54.
123 The problem is often claimed to have its origins in discussion in D. Parfit's *Reasons and Persons*, Oxford: Oxford University Press, 1986, p. 359. However, there are

presentations of it in the literature that pre-date this. See, for example, M. Bayles, 'Harm to the Unconceived', *Philosophy and Public Affairs* 5, 1976, 292–304, or G. Kavka, 'The Paradox of Future Individuals', *Philosophy and Public Affairs* 11, 1982, 93–112. Because the problem has acquired such a life of its own I do not present it in Parfit's, or anyone else's, particular terms.

124 As Jeffrey Reiman puts it, '[T]here are no individuals apart from the lives they live; none that are waiting in the wings, so to speak, to live this or that life,' J. Reiman, 'Being Fair to Future People: The Non-Identity Problem in the Original Position', *Philosophy and Public Affairs* 35, 2007, 69–92, p. 70.

125 As Reiman puts it, 'particular individuals only care about existing as those particulars once they are alive', ibid., p. 87.

126 Ibid., p. 85.

127 Ibid., p. 80, emphasis in original.

128 Rawls, *A Theory of Justice*, rev. edn, p. 8. See also Rawls, *Justice as Fairness*, p. 13.

129 Rawls, *A Theory of Justice*, rev. edn, p. 8.

130 Ibid., p. 8.

131 See, for example, C. Farrelly, 'Justice in Ideal Theory: A Refutation', *Political Studies* 55, 2007, 844–64.

132 A. Swift, 'The Value of Philosophy in Nonideal Circumstances', *Social Theory and Practice* 34(3), 2008, 363–87, p. 382. See also his excellent discussion of the hazy distinction between 'epistemological' (roughly, knowledge-seeking) and 'practical' (roughly, action-guiding) conceptions of political philosophy, ibid., pp. 366–68.

133 I. Robeyns, 'Ideal Theory in Theory and Practice', *Social Theory and Practice*, 34(3), 2008, 341–62, p. 362.

134 Ibid., p. 347.

135 Ibid., pp. 348–52.

136 Ibid., p. 360.

137 Rawls, *Political Liberalism*, p. 284.

138 See ibid., p. 285. See also M. Ronzoni, 'The Global Order: A Case of Background Injustice? A Practice-Dependent Account', *Philosophy and Public Affairs* 37(3), 2009, 229–56, for an excellent discussion of the much misunderstood relationship between the basic structure and background justice in debates about global justice.

139 See Rawls, *Political Liberalism*, p. 266.

3 Precautions

1 F. Pearce, *With Speed and Violence: Why Scientists Fear Tipping Points in Climate Change*, Boston: Beacon Press, 2007.

2 See J. Houghton, *Global Warming: The Complete Briefing*, Cambridge: Cambridge University Press 2009, Chapter 9 for detailed discussion of the nature and types of scientific uncertainty with respect to climate change.

3 I borrow this term from the excellent discussion in S. Gardiner, 'Saved by Disaster? Abrupt Climate Change, Political Inertia and the Possibility of an Intergenerational Arms Race', *Journal of Social Philosophy*, Vol. XL, No. 2, 2009, 140–62.

4 The IPCC is, at the time of writing, in the process of preparing a Special Report on 'Extreme Events and Disasters: Managing the Risks'. The Scoping Paper for the Special Report is available at: http://ipcc-wg2.gov/AR5/extremes-sr/extremes_documents/SREX-Oslo_Proceedings.pdf (accessed 27 July 2010).

5 J. Hansen, 'Tipping Point: Perspective of a Climatologist: Abstract', http://pubs.giss.nasa.gov/abstracts/inpress/Hansen_1.html (accessed 5 July 2007).

6 'Warming Hits Tipping Point', *The Guardian*, 11 August 2005, available at: http://www.guardian.co.uk/climatechange/story/0,12374,1546824,00.html (accessed 27 July 2010).

7 See Hansen, 'The Earth's Energy Imbalance: Confirmation and Implications', *Science*, Vol. 308, 1431–35. Available for download at: http://pubs.giss.nasa.gov/authors/jhansen.html (accessed 27 July 2010).
8 See Pearce, *With Speed and Violence*, pp. 152f.
9 Pearce, *With Speed and Violence*, p. 150.
10 See Pearce, *With Speed and Violence*, pp. 90–98.
11 See M. Lynas, *Six Degrees: Our Future on a Hotter Planet*, London: HarperCollins, 2007, pp. 207–62.
12 See J. Rawls, *A Theory of Justice* (rev. edn). Oxford: Oxford University Press, 1999, pp. 251–58.
13 See Rawls, *A Theory of Justice* (rev. edn), p. 121.
14 Cf. K.S. Ekeli, 'Environmental Risks, Uncertainty and Intergenerational Ethics', *Environmental Values* 13(4), 2004: 426–27.
15 See C. Spash, *Greenhouse Economics: Value and Ethics* London: Routledge, 2002, p. 122.
16 See, for example, R. Posner, *Catastrophe: Risk and Response*, Oxford: Oxford University Press, 2004, p. 148; C. Sunstein, 'The Catastrophic Harm Precautionary Principle', *Issues in Legal Scholarship* 7, 2007, Article 3, http://www.bepress.com/ils/iss10/art3 (accessed 27 July 2010); C. Sunstein, *Worst Case Scenarios*, Harvard: Harvard University Press, 2007, chapter 3.
17 For example, see S.O. Hansson, 'The Limits of Precaution', *Foundations of Science*, 2(2), 1997, 293–306; P. Sandin, 'The Precautionary Principle and the Concept of Precaution', *Environmental Values*, 13, 2004, 461–75; European Environment Agency, *The Precautionary Principle in the 20th Century*, London: Earthscan, 2002; D. Turner and L. Hartzell, 'The Lack of Clarity in the Precautionary Principle', *Environmental Values* 13, 2004, 449–60; R.S. Dimitrov, 'Precaution in Global Environmental Politics', *International Journal of Global Environmental Issues*, 5(1/2), 2005, 96–113.
18 Emphasis added.
19 Emphasis added.
20 C. Sunstein, *Laws of Fear: Beyond the Precautionary Principle*, Cambridge: Cambridge University Press, 2005, p. 14.
21 Note, though, that Sunstein does defend a precautionary principle for uncertain catastrophes, the paradigm of which he takes to be climate change catastrophes; Sunstein, 'The Catastrophic Harm Precautionary Principle'.
22 Sunstein, *Worst Case Scenarios*, p. 14.
23 A classic discussion of the perception of risk is P. Slovic, 'Perception of Risk', *Science*, New Series 236(4799) 1987, 280–85.
24 For discussion see J. Lemons, K. Shrader-Frechette and C. Cranor, 'The Precautionary Principle: Scientific Uncertainty and Type I and Type II Errors', *Foundations of Science* 2(2), 1997, 207–36.
25 As Harremoës et al. point out, there are lots of examples of false negatives related to processes and activities which turned out to be harmful but which were preventable through precautionary measures – indeed, their book contains fourteen case studies that illustrate this. However, there are very few false positives related to precautionary activities which turned out to have been unnecessary because the process or activity regulated with precautionary measures turned out to be completely harmless; their example is of the Y2K bug. See Harremoës et al., *The Precautionary Principle in the 20th Century*, London: Earthscan, 2002, p. 3.
26 See ibid., p. 191.
27 For a comprehensive list of international legal documents in which the Weak Precautionary Principle and the Strong Precautionary Principle appear see European Environment Agency, ibid., p. 6.
28 For more on the weak precautionary principle see E. Soule, 'Assessing the Precautionary Principle', *Public Affairs Quarterly* 14(4), 2000, 309–28.

29 This could be either a consequence of the nature of the principle (it is non-categorical), or a consequence of the unprincipled nature of policy makers, or both. I take no stand here on the question of whether a principle must be categorical to count as such, or to qualify as adequate. As for policy makers, it is almost certainly the case that they are often motivated by reasons not justified by good principles.
30 On the latter accusation see, for example, Sunstein, *Worst Case Scenarios*, chapter 1.
31 A.M. Freeman III, 'Economics' in D. Jamieson (ed.), *A Companion to Environmental Philosophy*, Oxford: Blackwell, 2003, p. 281.
32 For a critique of cost-benefit analysis along these lines see E. Anderson, *Value in Ethics and Economics*, Cambridge, MA: Harvard University Press, 1993, chapter 9.
33 R. Posner, *Catastrophe: Risk and Response*, Oxford: Oxford University Press, 2004, p. 148.
34 C. Sunstein, *Laws of Fear: Beyond the Precautionary Principle*, Cambridge: Cambridge University Press, 2005, p. 14.
35 Sunstein, *Worst Case Scenarios*, p. 14.
36 See ibid., p. 168.
37 See ibid., pp. 167–68.
38 See ibid., p. 173, in particular.
39 See Rawls, *Justice as Fairness*, pp. 94–96.
40 See Sunstein, *Worst Case Scenarios*, p. 171.
41 See, for example, C. Sunstein, 'Cost-Benefit Analysis and the Environment', John M. Olin Law and Economics Working Paper No. 227 (2D Series), October 2004, The Law School, University of Chicago. Available on line at http://www.law.uchicago.edu/files/files/227-crs-environment.pdf (accessed 27 July 2010). Sunstein offers the most persuasive defence available of the U.S Environmental Protection Agency's notorious valuation of a human life at $6.1 million (which is not, of course, to say that it persuades).
42 As already noted, this argument does not turn on the controversial claim that minimising the risks of any worst-case scenario is the principle of rational choice in all circumstances of uncertainty. Rather, it is (a) limited to climate change catastrophes and (b) an argument from egalitarian justice, which I do not take to be justified by reference to principles of rational choice.
43 Both figures given in G. Monbiot, *Heat: How To Stop the Planet Burning*, London: Allen Lane, 2006, p. 51.
44 With a range of -1 per cent to +3.5 per cent. *The Stern Review*, 2006, Executive Summary, pp. xiii–xvi.
45 See Datablog, *Guardian*, 'UK Bank Bailouts in Full. What Else Could You Get for £38bn?'. Online. HTTP: <http://www.guardian.co.uk/news/datablog/2009/nov/03/bank-bailouts-uk-credit-crunch#data> (accessed 19 August 2010).
46 It is worth noting that the content of IPCC Reports is decided by consensus among the authors. Given their often deep divergences of scientific opinion, the net effect of this method is to produce conservative statements – the lowest common denominator of scientific opinion to which all authors can agree.
47 Note that Gardiner refers to this principle as the 'Rawlsian Core Precautionary Principle'. S. Gardiner, 'A Core Precautionary Principle', *The Journal of Political Philosophy* 14(1), 2006, 33–60, pp. 47–48.
48 See ibid., p. 47. Gardiner also adds a criterion that specifies that the threats addressed by any precautionary principle must be 'realistic' (p. 51), e.g., not science fiction, purely imagined, paranoid, purely religiously inspired, etc. Note that the science that informs Methane Nightmare is sound.
49 For discussion see Gardiner, 'A Core Precautionary Principle', pp. 56–57.
50 Note, however, that Rawls himself does not conceive of his principle of intergenerational justice (the just savings principle) as a principle of distributive justice. For an interpretation of the just savings principle as a principle of distributive justice see

F. Gaspart and A. Gosseries, 'Are Generational Savings Unjust? *Politics, Philosophy and Economics* 6, 2007, 193–217.

51 See for example B. Lomborg, 'Climate Change Can Wait. World Health Can't', *Observer*, 2 July 2006.

52 See, for example, F. Ackerman and L. Heinzerling, *Priceless: On Knowing the Price of Everything and the Value of Nothing*, New York: The New Press, 1994; E. Anderson, *Value in Ethics and Economics*, Cambridge, MA: Harvard University Press, 1993.

53 Of course, economists and utilitarians – for whom all values are commensurable – will vigorously object at this point. Classic defences of the existence of incommensurable values fit to meet these objection can be found in (for example), I. Berlin, 'The Pursuit of the Ideal', *The Crooked Timber of Humanity*, London: John Murray, 1990; J. Raz, *The Morality of Freedom*, Oxford: Clarendon Press, 1986.

54 Again, economists and utilitarians committed to denying the existence of incommensurable values will not be moved by such an argument, as they will reject the incommensurable value of intergenerational justice that lies at its heart. However, it is important to note that, in this case, all roads could lead to Rome. Economists and consequentialists can make alternative arguments for the strong precautionary principle with respect to climate change catastrophes that do not depend on the incommensurable value of doing intergenerational justice. See, for example, Posner, *Catastrophe*, chapter 3; Sunstein, 'The Catastrophic Harm Precautionary Principle'.

55 J. Rawls, *Justice as Fairness: A Restatement*, Cambridge, MA: Harvard University Press, 2001, p. 102.

56 Ibid., p. 128.

57 See G. Dickens, 'The Blast in the Past', *Nature* 401, 1999, 752–55.

58 S.L. Wing et al., 'Transient Floral Change and Rapid Global Warming at the Paleocene-Eocene Boundary', *Science* 310, 2005, 333–36.

59 See S. Manabe et al., 'Century-scale Change in Water Availability: CO2 Quadrupling Experiment', *Climatic Change* 64, 2004, 59–76.

60 Chris Huntingford et al., 'Regional Climate-Model Predictions of Extreme Rainfall for a Changing Climate', *Quarterly Journal of the Royal Meteorological Society* 129, 2003, 1607–21.

61 See Lynas, *Six Degrees*, pp. 223–30. A compelling vision of what it might be like to try to survive in such a world is found in C. McCarthy's novel *The Road*, Random House, New York: 2006.

62 See Lynas, *Six Degrees*, pp. 233ff.

63 See D. Kidder and T. Worsley, 'Causes and Consequences of Extreme Permo-Triassic Warming to Globally Equitable Climate and Relation to the Permo-Triassic Extinction and Recovery', *Palaeogeography, Palaeoclimatology, Palaeoecology* 203(3/4), 2004, 207–37.

64 See G. Ryskin, 'Methane-Driven Oceanic Eruptions and Mass Extinctions', *Geology* 31(9), 2003, 741–44.

65 See J.K. Nielsen and Y. Shen, 'Evidence for Sulphidic Deep Water During the Late Permian in the East Greenland Basin', *Geology* 32(12), 2004, 1037–40.

66 See L.R. Kump et al., 'Massive Release of Hydrogen Sulfide to the Surface Ocean and Atmosphere During Intervals of Ocean Anoxia', *Geology* 33(5), 2005, 367–400.

67 Lynas, *Six Degrees*, p. 254, emphasis in original.

68 Note that this is not to say that there could be no justice between people in such circumstances, or between us and them; rather, it is to say that they would be in no position to achieve justice. See B. Barry, 'Circumstances of Justice and Future Generations' in R.I. Sikora and B. Barry (eds), *Obligations to Future Generations*, Philadelphia: Temple University Press, 1978, pp. 204–48.

69 For reflections bearing on this point see J.M. Taurek, 'Should the Numbers Count?', *Philosophy and Public Affairs* 6(4), 1977, 293–316.

70 I grant that this claim does not hold in all possible worlds: for example, when the costs of precautions are 100 per cent of GDP every year then the harms of Unnecessary Expenditure could exceed the harms of Methane Nightmare. However, this is not the case, even according to the highest estimates of the costs of precautions. That the claim is true in the actual world is good enough for me.
71 For discussion of whether placing the strains of commitment at the heart of political justification in this way is consistent with a contractualist approach see J. Hampton, 'Contracts and Choices: Does Rawls Have a Social Contract Theory?', *The Journal of Philosophy* 77(6), 1980, 315–38.
72 Although note my comments in Chapter 2 about how science can sometimes convert states of knowledge into states of uncertainty: climate change science is not guaranteed to shrink the set of propositions concerning climate change with respect to which we are uncertain.
73 See Sunstein, *Worst Case Scenarios*, pp. 159–62.
74 Thanks are due to Jo Wolff for pressing this objection with me.
75 Mike Otsuka made the point to me in this way, although I think the addition of a big party is mine.
76 I am grateful to Jurgen de Wispelaere for suggesting this way forward to me.
77 See Posner, *Catastrophe*, pp. 24–29.
78 Perhaps the strong precautionary principle would be insulated to some degree against the priorities problem by incorporating into the account Gardiner's 'realistic outcomes' criterion (see note 48). This would, for example, probably rule out alien invasions as events against which climate change catastrophes compete for resources necessary for precautionary action. However, it would not rule out everything: comet collisions and supervolcanic explosions are very real possibilities.
79 This principle states that a person ought to be free from interference by the state and by others unless her actions affect the interests of others and cause them harm. See J.S. Mill, *On Liberty*, Cambridge: Cambridge University Press, 1989, chapter 1.
80 For a classic discussion see B. Williams, 'Moral Luck' in his *Moral Luck*, Cambridge: Cambridge University Press, 1981, pp. 20–40.
81 For discussions that bear on this hard question see J.J. Thomson, 'Remarks on Causation and Liability', *Philosophy and Public Affairs* 13(2), 1984, 101–33.
82 Hansson, 'The Limits of Precaution', p. 300.
83 For reasons of style I shall suppress the qualifier 'anthropogenically caused' for the remainder of this section.
84 Spash, *Greenhouse Economics*, p. 140.
85 See European Environment Agency, *The Precautionary Principle in the 20th Century*, pp. xv–xvi.
86 IPCC, 'Guidance Notes for Lead Authors of the IPCC Fourth Assessment Report on Addressing Uncertainties', p. 3.
87 The IPCC actually refers to its qualitative representation as 'levels of understanding'; however, this makes no difference for my purposes. See ibid., p. 3.
88 See ibid., p. 3.
89 See, for example, M. Manning et al., 'Describing Scientific Uncertainties in Climate Change to Support Analysis of Risk and of Options', Report on IPCC Workshop on Uncertainty and Risk, 2004, p. 25.
90 See M. Manning and M. Petit, 'A Concept Paper for the AR4 Cross Cutting Theme: Uncertainties and Risk', p. 6. For a review of IPCC practice with respect to uncertainty see M. Ha-Duong, R. Swart and L. Bernstein, 'Uncertainty Management in the IPCC', *Global Environmental Change*, 2007, p. 17. It is worth noting that the IPCC appears not to have successfully communicated the distinction between levels of confidence and assignments of probability to the general public: one study has shown that statements assigned a low level of confidence are

understood thereby to have a low level of probability. See D.V. Budescu, S. Broomell and H. Por, 'Improving Communication of Uncertainty in the Reports of the Intergovernmental Panel on Climate Change', *Psychological Science* 20(3), 2009, 299–308.

91 The point made in this paragraph should be taken as an attempt to satisfy the European Environment Agency's 'lessons' 8 and 9 that application of the precautionary principle should '[e]nsure use of "lay" and local knowledge, as well as relevant specialist expertise … ' and '[t]ake full account of the assumptions and values of different social groups'. European Environment Agency, *The Precautionary Principle in the 20th Century*, p. 186.

92 Witness how erstwhile creationists have recast themselves as advocates of 'intelligent design'.

93 See, for example, Posner, *Catastrophe*, chapter 3; Sunstein, 'The Catastrophic Harm Precautionary Principle'.

94 See Sunstein, *Laws of Fear*, p. 2.

4 Intergenerational corrective justice

1 The most thorough explorations to date of the role and nature of principles of distributive justice in thinking about climate change politics can be found in the work of S. Caney, especially S. Caney, 'Cosmopolitan Justice, Responsibility and Global Climate Change', *Leiden Journal of International Law* 18(4), 2005, 747–5, and S. Caney, 'Justice and the Distribution of Greenhouse Gas Emissions', *Journal of Global Ethics* 5(2), 2009, 125–46.

2 A generation is here understood in familial terms as spanning roughly 25 years. Thus, six generations hence is equivalent to 150 years hence. The atmospheric life of CO_2 is 50–200 years. Thus, even if we made drastic cuts to our CO_2 emissions now, members of up to and including 6–8 generations hence would probably still experience the harmful effects of our excessive emissions pre-cuts.

3 For an excellent treatment of corrective, retributive and distributive justice from a broadly Rawlsian point of view see A. Ripstein, *Equality, Responsibility and the Law*, Cambridge: Cambridge University Press, 1999.

4 The most prominent of this group is Bjørn Lomborg. See B. Lomborg, *The Skeptical Environmentalist: Measuring the Real State of the World*, Cambridge: Cambridge University Press, 2001.

5 See, for example, D.A. Farber, 'Basic Compensation for Victims of Climate Change', *University of Pennsylvania Law Review* 155, 2007, 1605–56; M.D. Adler, 'Corrective Justice and Liability for Global Warming', *University of Pennsylvania Law Review* 155(6), 2007, 1859–67.

6 J. Coleman, 'The Practice of Corrective Justice' in D.G. Owen (ed.), *Philosophical Foundations of Tort Law*, Oxford: Clarendon Press, 1995, esp. pp. 66–67.

7 For more on the distinction between distributive and corrective justice see E.J. Weinrib, 'Correlativity, Personality, and the Emerging Consensus on Corrective Justice', *Theoretical Inquiries in Law (Online Edition)* 2(1), 2001, Article 4. Online. HTTP: <http://www.bepress.com/til/default/vol2/iss1/art4> (accessed 10 August 2010).

8 This is not, of course, the only – or even the most popular – theoretical approach to tort law. Its main competitor is the economic approach, whereby tort law is analysed in terms of the efficient distribution of the costs of harms caused to victims by tortious actions. The classic defence is in G. Calabresi, *The Costs of Accidents: A Legal and Economic Analysis*, New Haven: Yale University Press, 1970. For a review of the emerging literature on tort law and corrective justice see D.G. Owen, 'Why Philosophy Matters to Tort Law' in D.G. Owen (ed.), *Philosophical Foundations of Tort Law*, Oxford, Clarendon Press, 1995. See also, for example, Weinrib, 'Correlativity,

Personality, and the Emerging Consensus on Corrective Justice'; S. Perry, 'The Mixed Conception of Corrective Justice', *Harvard Journal of Law and Public Policy* 15(3), 1992, 917–39.

9 A. Ripstein, 'Philosophy of Tort Law' in J.L. Coleman and S. Shapiro (eds), *The Oxford Handbook of Jurisprudence and Legal Philosophy*, Oxford: Oxford University Press, 2001, p. 656.

10 See, for example, ibid.

11 See, for example, J. Coleman, *Risks and Wrongs*, Cambridge: Cambridge University Press.

12 See, for example, Weinrib, 'Correlativity, Personality, and the Emerging Consensus on Corrective Justice'.

13 In particular, C. Schroeder, 'Corrective Justice and Liability for Increasing Risks', *UCLA Law Review* 37, 1989–90, 439–78. See also C. Schroeder, 'Corrective Justice, Liability for Risks, and Tort Law', *UCLA Law Review* 38, 1990–99, 143–61.

14 Schroeder, 'Corrective Justice and Liability for Increasing Risks', p. 449. However, compare with Ripstein, who argues that the notion of responsibility operative in tort law qua expression of corrective justice is entirely independent of moral responsibility as commonly described in the language of blame. Instead, on his view, this idea of responsibility captures the restrictions on their own and others' behaviour that a reasonable person concerned to balance important interests in liberty and security would accept. See, for example, Ripstein, *Equality, Responsibility and the Law*, p. 289.

15 Note that this condition accommodates both fault and strict liability when they are distinguished, as they should be, just in terms of the content of the duty breached.

16 Schroeder, 'Corrective Justice and Liability for Increasing Risks', p. 450. The labels in square brackets are Schroeder's own.

17 The other part of the causation requirement is that the party is the proximate cause of the harm, which is often interpreted as the requirement that the effects of the injurer's actions were foreseeable.

18 Schroeder, 'Corrective Justice and Liability for Increasing Risks', pp. 455–56.

19 See C. Schroeder, 'Causation, Compensation, and Moral Responsibility' in David G. Owen (ed.), *Philosophical Foundations of Tort Law*, Oxford: Clarendon Press, 1995, pp. 347–62.

20 See J. Waldron, 'Moments of Carelessness and Massive Loss' in J. Waldron, *Liberal Rights: Collected Papers 1981–1991*, Cambridge: Cambridge University Press, 1993, p. 387. Jules Coleman describes and discusses an almost identical example in Coleman, *Risks and Wrongs*, pp. 321f.

21 Waldron, 'Moments of Carelessness and Massive Loss', p. 399, italics in original. See also Coleman, *Risks and Wrongs*, p. 274.

22 Note that there are some cases where liability attaches to persons who could not reasonably have been expected to know that their actions would cause harm. For fascinating and important reflections on the 'culpable ignorance' that characterises these cases see I. Hacking, 'Culpable Ignorance of Interference Effects' in D. MacLean (ed.), *Values at Risk*, New Jersey: Rowman and Allanheld, 1986.

23 The *ex ante* conception of responsibility can ground ascriptions of both fault and strict liability. I shall have more to say about these different forms of liability in the next section.

24 Schroeder, 'Corrective Justice and Liability for Increasing Risks', p. 467.

25 This marks a difference between the account given here and that given by David McCarthy. He denies that harm is a necessary condition for liability, which means that the imposition of risk on B by A *in itself* can ground B's claim for compensation, even if the harm never materialises. See D. McCarthy, 'Liability and Risk', *Philosophy and Public Affairs* 25(3), 1996, 238–62, especially the discussion on pp. 247–49. In contrast, what I deny here is the requirement of cause-in-fact for agent liability, *not* the requirement that harm be suffered by a person for a compensation claim

correlating with that liability to be rightful. It may be that the person put at risk by others ought to have legal redress against them prior to suffering harm when the risk ripens, or even if the risk never ripens, but I do not think an account of liability based on corrective justice will support such an argument, and certainly not as an aspect of tort law.

26 Judith Jarvis Thomson calls such cases instances of 'pure risk imposition'. The modified account of liability outlined here incorporates this type of risk-imposition in addition to (less controversially) 'impure risk imposition', whereby 'an agent causes an unwanted outcome, and imposes a risk of a further unwanted outcome', J.J. Thomson, 'Imposing Risks' in J.J. Thomson, *Rights, Restitution and Risk*, Cambridge, MA: Harvard University Press, 1986, p. 173. Cf. also S. Vanderheiden, *Atmospheric Justice*, Oxford: Oxford University Press, 2008, who makes a similar point with respect to negligence.

27 Schroeder, 'Corrective Justice and Liability for Increasing Risks', p. 440.

28 Ibid., p. 440.

29 Ibid., p. 468.

30 Coleman, *Risks and Wrongs*, p. 286.

31 For discussion of this response in the context of moral responsibility see T. Nagel, 'Moral Luck' in his *Mortal Questions*, Cambridge: Cambridge University Press, 1979, pp. 35–36.

32 See also Waldron, 'Moments of Carelessness and Massive Loss', p. 398–99.

33 Rawls claims that 'reasonable pluralism' must be assumed as a permanent part of the context in which principles of justice are to operate. A large part of what it means for pluralism to be reasonable is for people to accept that the exercise of human reason in conditions of freedom will inevitably lead to great disagreement in matters of judgement (such as debates about free will and responsibility), in virtue of the 'burdens' the exercise of judgement is subject to (for example, the complexity of conflicting bodies of evidence supporting opposed views, or the effect of each person's different life story on what they come to believe). See J. Rawls, *Political Liberalism*, New York: Columbia University Press, 1993, pp. 54–58, and pp. 36f.

34 Cf. Schroeder, 'Corrective Justice and Liability for Increasing Risks', p. 474, whose comments here about risk aversion inspired the arguments presented in (a) and (b). See also Jeremy Waldron, who claims with respect to no-fault schemes of compensation such as the NZ scheme, 'if there was ever a case for maximin, this is it', Waldron, 'Moments of Carelessness and Massive Loss', p. 408.

35 See Schroeder, 'Corrective Justice and Liability for Increasing Risks', pp. 456–57.

36 It is important to bear in mind here that my arguments are limited to principles fit to inform private rather than the criminal law. It might be that the arguments could be extended to criminal liability, but I do not commit to this claim, nor explore it here. Insightful discussion can be found in Coleman, *Risks and Wrongs*, pp. 222–26.

37 For an alternative view – whereby holding Fate liable for his bad luck in hitting Hurt involves no unfairness to him because the luck in question is optional – see M. Otsuka, 'Moral Luck: Optional not Brute', *Philosophical Perspectives*, 23, 2009, 373–88.

38 I think that these assumptions inform some of McCarthy's arguments for treating risk-imposition as a ground for liability and compensation in some cases. In all the cases he discusses the person on whom a risk is imposed, and who ipso facto has a claim for compensation against the risk-imposer, is aware of this fact – indeed, they deliberate in some detail about what the risk means to them. See McCarthy, 'Liability and Risk', pp. 247–49.

39 For further arguments against the thesis that risks are harms see S. Perry, 'Risk, Harm, Interests and Rights' in T. Lewens (ed.), *Risk: Philosophical Perspectives*, London: Routledge, 2007. For contrary argument that (some) risks are harms

see C. Finkelstein, 'Is Risk a Harm?', *University of Pennsylvania Law Review* 151, 2003, 963–1001.

40 Thomson, 'Imposing Risks', p. 180, emphasis in original.

41 On these sorts of claims, see Coleman, *Risks and Wrongs*, pp. 274–75. He argues that what matters for responsibility – and liability – is that an agent is the author of a wrongful act or form of behaviour, and that this can be true of an agent who wrongfully omits to do something despite the fact that her omission is not the cause of a harm suffered by another that could have been prevented by avoiding the wrongful omission: 'Liability may require responsibility, but responsibility does not require causation' (ibid., p. 274).

42 Thomson, 'Imposing Risks', p. 185.

43 For discussion see J.J. Thomson, 'Self-Defense and Rights', in J.J. Thomson, *Rights, Restitution and Risk: Essays in Moral Theory*, Cambridge, MA: Harvard University Press, 1986.

44 Thomson, 'Imposing Risks', p. 187.

45 Thomson's objection to this strategy consists of a set of questions (left unanswered) about the consequences of abandoning the inheritance principle in question for the role of other inheritance principles in the moral assessment of acts. She asks:

> [A]re we to say that all inheritance principles of the … form …
>
> If A ought not to VP1, then if it is the case that if A VP2s, he will thereby VP1, then A ought not to VP2, are false? Or only some subset of them? And if all of them are false, are there any inheritance principles of any other form which are true? Is there no 'nesting' of impermissibilities? Does every set of ascriptions of impermissibility such that no one of them entails any other form a mere concatenation – each requiring independent justification? (ibid., p. 186).

Perhaps we are to say this. In any case, it is not clear to me that these big questions must be settled in order for the claim I make in the text to hold. This is that the following inheritance principle ought to be abandoned: If the current generation ought not to cause climate change harms to future generations, then if it is the case that if the current generation does not drastically reduce its emission of greenhouse gases, it will thereby cause harm to future generations, then the current generation ought not to fail to drastically reduce its emissions of greenhouse gases. My claim is that we do not need a certain link between the current generation's failure to reduce greenhouse gas emissions and the causation of harm to future generations in order to require that the current generation make cuts to its greenhouse gas emissions, and I provided argument for the independent claim in the last chapter.

46 The UK Criminal Injuries Compensation Authority exemplifies this commitment: it pays compensation to victims of violent crime even if the criminals are not prosecuted.

47 For an account of liability which gives such a central role to foresight see Vanderheiden, *Atmospheric Justice*, pp. 132–38.

48 Here I follow Jules Coleman in distinguishing between fault and strict liability by reference to the content of the underlying duty of care that is breached in each case, rather than by reference to the necessity of establishing faulty agency for fault liability, but not for strict liability. See J. Coleman, 'Theories of Tort Law' in E.N. Zalta (ed.), *The Stanford Encyclopedia of Philosophy*, 2003. Online. HTTP: <http://plato.stanford.edu/entries/tort-theories/> (accessed 10 August 2010).

49 Coleman, *Risks and Wrongs*, p. 220.

50 Gerald Postema expresses scepticism about the possibility of fault in the doing without fault in the doer. He argues that 'there are no actions without agents, and thus we speak of faulty actions only derivatively. Actions may be good or bad, right or wrong, but only agents bear responsibility for them and they alone are bearers of

fault.' G.J. Postema, 'Risks, Wrongs, and Responsibility: Coleman's Liberal Theory of Commutative Justice', *Yale Law Journal* 103, 1993–94, 861–97, pp. 895–96. I return to this point below.

51 Coleman, *Risks and Wrongs*, pp. 219–20.

52 Ibid., p. 228.

53 J. Coleman, 'The Morality of Strict Tort Liability', *William and Mary Law Review* 18, 1976, 259–86.

54 The labels for these categories in parentheses are taken from Coleman, *Risks and Wrongs*, p. 260.

55 In this I disagree with Vanderheiden's blanket account of legal responsibility which, he claims, 'depends on its correspondence with judgments of moral responsibility for its legitimacy' (Vanderheiden, *Atmospheric Justice*, p. 150); whereas this is true of criminal responsibility, it is not true of tort responsibility. Throughout his discussion of moral responsibility and greenhouse gas emissions Vanderheiden rolls together criminal and tort liability in a way that obscures the nuances of the latter; see, especially, ibid., pp. 158–59 and p. 171.

56 For discussion of justifications and excuses in the criminal law, see M. Baron, 'Justifications and Excuses', *Ohio State Journal of Criminal Law* 2, 2005, 387–406.

57 McCarthy, 'Liability and Risk', p. 256.

58 As some commentators have noted, a just pattern of distribution provides the background against which further just transactions and relationships take place: if recovery of damages by a victim disrupts this background distribution, no defender of a patterned theory of distribution would insist that the victim's award of compensation be taxed at 100 per cent and redistributed to the injurer so as to reinstitute the original pattern. See Ripstein, *Equality, Responsibility and the Law*, p. 28; Waldron, 'Moments of Carelessness and Massive Loss', p. 394; Coleman, *Risks and Wrongs*, p. 305, pp. 311f, pp. 350–54.

59 See Coleman, *Risks and Wrongs*, pp. 326–28, and chapters 18 and 19.

60 Bono flew his favourite trilby hat first class from London to Modena in order to wear it while performing at a concert with Pavarotti. See 'Meet the Kings (and Queens) of excess', *Mail Online*, 16 June 2006. Online. HTTP: <http://www.dailymail.co.uk/tvshowbiz/article-390967/Meet-Kings-Queens-excess.html>.

61 See E. Rosenthal, 'Third-World Stove Soot Is Target in Climate Fight', *New York Times*, 15 April 2009. Online. HTTP: <http://www.nytimes.com/2009/04/16/science/earth/16degrees.html> (accessed 13 August 2010).

62 See B. Law, 'Can Palm Oil Help Indonesia's Poor?', *Guardian*, 1 March 2010. Online. HTTP: <http://news.bbc.co.uk/1/hi/world/asia-pacific/8534031.stm> (accessed 12 August 2010). Indonesia is deforesting and draining its peatlands at an alarming rate – 10 million of its 22.5 million hectares have disappeared already – to make way for palm oil plantations to feed a growing demand for biofuels in Western countries. Indonesian peatlands cover 0.1 per cent of the Earth's surface but are responsible for 4 per cent of global emissions of greenhouse gases. See Greenpeace UK, 'Palm Oil'. Online. HTTP: <http://www.greenpeace.org.uk/forests/palm-oil> (accessed 12 August 2010).

63 Coleman, *Risks and Wrongs*, p. 260.

64 Ibid., p. 260.

65 H. Shue, 'Subsistence Emissions and Luxury Emissions', *Law and Policy* 15(1), 1993, 39–59, p. 55.

66 Ibid., p. 56. See also H. Shue, 'Climate' in D. Jamieson (ed.), *A Companion to Environmental Philosophy*, Oxford: Blackwell, 2003.

67 Ibid., p. 451.

68 Almas caviar is produced in Iran and costs £10,000–£18,000 per kilo.

69 A dark possibility on this approach is that the allocation that would result would be insufficient for anyone's survival. If this is the reality, then, arguably, principles of

distributive justice no longer apply, and principles of triage ought to be adopted. The relationship between the two sets of principles is murky at best. I take up this topic in Chapter 5.

70 Cf. the discussion in Vanderheiden, *Atmospheric Justice*, pp. 163–67.

71 D.A. Farber, 'The Case for Climate Compensation: Doing Justice to Climate Change Victims in a Complex World', *Utah Law Review* 2, 2008, 1–36, 20–21.

72 In 2008 total emissions of CO_2 (million metric tons) for China was 6533.554, and for the US was 5832.818. However, the discrepancy between them in terms of per capita CO_2 emissions remains striking: in 2008 China emitted 4.91228 metric tons of CO_2 per person, whereas the US emitted 19.18313 metric tons of CO_2 per person. Figures from the US Energy Information Administration. Online. HTTP: <http://tonto.eia.doe.gov/cfapps/ipdbproject/iedindex3.cfm?tid=90&pid=45&aid=8&cid=& syid=2004&eyid=2008&unit=MMTCD> (accessed 12 August 2010).

73 E. Posner and C.R. Sunstein, 'Climate Change Justice', John M. Olin Law and Economics Working Paper No. 354 (2D series), Public Law and Legal Theory Working Paper No. 177. Online. HTTP: <http://papers.ssrn.com/sol3/papers.cfm?abstract_ id=1008958> (accessed 10 August 2010), p. 28.

74 See Farber, 'The Case for Climate Compensation: Doing Justice to Climate Change Victims in a Complex World', pp. 14f.

75 Ibid., p. 17.

76 Farber makes a further counterargument to Posner and Sunstein on the grounds that the causal overdetermination of climate change harms – to which the saturation effect they posit is tantamount – does not relieve the US of liability for these harms, even granting their assumption that there is a saturation effect. Causal overdetermination does not mean that any, or all, of the causes which overdetermine are not in fact causes. See ibid., pp. 14–16.

77 The US Court of Appeals for the Fifth Circuit ruled (16 October 2009) that plaintiffs had standing to proceed with allegations of nuisance, negligence, and trespass.

78 For discussion of this, and other, cases, and of more general issues surrounding climate change litigation see D. Hunter and J. Salzman, 'Negligence in the Air: The Duty of Care in Climate Change Litigation', *University of Pennsylvania Law Review* 155(6), 2007, 1741–94.

79 Daniel Farber has consistently argued for such a fund in the intragenerational context. See Farber, 'Basic Compensation for Victims of Climate Change', esp. pp. 1607–8; Farber, 'The Case for Climate Compensation: Doing Justice to Climate Change Victims in a Complex World', esp. p. 31; D.A. Farber, 'Tort Law in the Era of Climate Change, Katrina and 9/11: Exploring Liability for Extraordinary Risks', *UC Berkeley Public Law Research Paper* No. 11211125, 2008. Online. HTTP: http://ssrn.com/abstract=1121125 (accessed 10 August 2010).

80 See the Accident Compensation Corporation website. Online. HTTP: <www.acc.co.nz> (accessed 12 August 2010).

5 The end of the world as we know it

1 See Met Office, 'Avoiding Dangerous Climate Change', Met Office, 2008. Online. HTTP: http://www.metoffice.gov.uk/publications/brochures/cop14.pdf> (accessed 10 August 2010), p. 12.

2 See S. Manabe, R.T. Wetherald, P.C.D. Milly, T.L. Delworth and R.J. Stouffer, 'Century-scale Change in Water Availability: CO_2 Quadrupling Experiment', *Climatic Change* 64, 2004, 59–76.

3 P. Makohon, 'Witness: Memoirs of the Famine of 1933 in Ukraine'. Online. HTTP: <http://faminegenocide.com/resources/witness/main.html> (accessed 27

July 2010). A prominent account of the famine and its causes is in R. Conquest, *The Harvest of Sorrow: Soviet Collectivization and the Terror-Famine*, Oxford: Oxford University Press,1986.

4 See C.A. Anderson, 'Heat and Violence', *Current Directions in Psychological Science* 10 (1), 2001, 33–38.

5 For a review of the literature bearing on psychological and emotional impacts of various aspects of the climate change problem see J. Swim, S. Clayton, T. Doherty, R. Gifford, G. Howard, J. Reser, P. Stern and E. Weber, *Psychology and Global Climate Change: Addressing a Multi-faceted Phenomenon and Set of Challenges. A Report by the American Psychological Association's Task Force on the Interface Between Psychology and Global Climate Change*, American Psychological Society, 2009. Online. HTTP: http://www.apa.org/science/about/publications/climate-change.pdf> (accessed 10 August 2010). It is also possible, however, that people in this world would exhibit what Kahn calls 'environmental generational amnesia', whereby they would treat their environmental experience as a baseline for comparison and as a result fail to acknowledge the extent to which their environment has deteriorated over generations (P. H. Kahn, *The human relationship with nature*. Cambridge, MA: MIT Press, 1999). This sad possibility is likely to generate apathy with respect to trying to improve the environment. However, it is arguable that the worse conditions get, the less sustainable such amnesia would become: it is hard to imagine, for example, that people experiencing global famine would believe that things were no better for their forbears.

6 Note that this is not tantamount to making claims about the direction such politics would take, which is beyond my (and perhaps anyone's) abilities. Nevertheless, if we are seriously failing to prevent future climate change catastrophes then the least we can do is to think about how to pursue politics in the midst of such catastrophes so that our descendants might, at some point, be in a position to make a contribution to the project of justice over generations more admirable than our own.

7 For a potted history see G.R. Winslow, *Triage and Justice*, Berkeley: University of California Press, 1982, chapter 1.

8 See J. Elster, *Local Justice: How Institutions Allocate Scarce Goods and Necessary Burdens*, New York: Russell Sage Foundation, 1992.

9 Of course, there is much debate about what constitutes a reasonable period of time, with parties in arguments over the allocation of scarce medical resources often divided along patient/physician lines: an extra six months of life may be a reasonable period from the first-person perspective of a patient who will otherwise die within a month, but may not count as such from the perspective of a physician directing the allocation of scarce and expensive drugs. I ignore this debate here.

10 H. Shue, 'Subsistence Emissions and Luxury Emissions', *Law and Policy* 15(1), 1993, 39–59.

11 For ease of exposition I shall in this section sometimes suppress the qualifier 'distributive': unless otherwise indicated, reference to justice in this section should be read as reference to distributive justice.

12 For a discussion of similar issues see B. Barry, 'Circumstances of Justice and Future Generations' in R.I. Sikora and B. Barry (eds), *Obligations to Future Generations*, Philadelphia: Temple University Press, 1978.

13 I take this term from A. Abizadeh, 'Cooperation, Pervasive Impact, and Coercion: On the Scope (not Site) of Distributive Justice', *Philosophy and Public Affairs* 35(4), 2007, 318–58, pp. 320f.

14 D. Hume, *Enquiries Concerning Human Understanding and Concerning the Principles of Morals*, Oxford: Clarendon Press, 1975, p. 186.

15 Note, however, that Hume does not characterize self-interest in these circumstances as rapacious, or morally condemn it in any way, and he is right not to do so. When

one's survival and that of one's children is at stake, to put one's own and their interests first is a deep human instinct not to be condemned, and is perhaps morally required.

16 J. Rawls, *Justice as Fairness: A Restatement*, Cambridge, MA: Harvard University Press, 2001, p. 50.

17 Ibid., p. 6.

18 Ibid., p. 10.

19 J. Rawls, *Political Liberalism*, New York: Columbia University Press, 1993, p. 266.

20 Ibid., p. 266.

21 Abizadeh, 'Cooperation, Pervasive Impact, and Coercion: On the Scope (not Site) of Distributive Justice', pp. 330–31, emphasis in original. Abizadeh's interpretation contrasts with that of Samuel Freeman, who treats existing social cooperation as setting the scope of distributive justice; Freeman's account of Rawls' understanding of distributive justice nevertheless remains illuminating. See S. Freeman, 'The Law of Peoples, Social Cooperation, and Human Rights' and 'Distributive Justice and the Law of Peoples', in S. Freeman, *Justice and the Social Contract*, Oxford: Oxford University Press, 2007.

22 Abizadeh, 'Cooperation, Pervasive Impact, and Coercion: On the Scope (not Site) of Distributive Justice', p. 334, emphasis added.

23 This distinction between the two conceptions of distributive justice is well made by Rawls. See e.g. Rawls, *Justice as Fairness*, pp. 50–51.

24 Possible exceptions are very self-destructive people whose consumption habits, lifestyle, or deliberate self-harm deplete the degree to which they possess this property.

25 For discussion see Winslow, *Triage and Justice*, p. 88.

26 See L. Temkin, *Inequality*, Oxford: Oxford University Press, 1993.

27 There are other such cases. For example, when resources sufficient to save one or more individuals within a category are also sufficient to save a higher number within that category. For instance, with the resources available we could give one person a reasonable period of time with quality of life, or ten people a reasonable period of time with quality of life. Call this the 'numbers' problem. For arguments bearing on this problem see J.M. Taurek, 'Should the Numbers Count?', *Philosophy and Public Affairs* 6(4), 1977, 293–316; D. Parfit, 'Innumerate Ethics', *Philosophy and Public Affairs* 7(4), 1978, 285–301.

28 Material in the following paragraphs is adapted from C. McKinnon, 'Cosmopolitan Hope' in H. Brighouse and G. Brock (eds), *The Political Philosophy of Cosmopolitanism*, Cambridge: Cambridge University Press, 2005.

29 For more on the senses of possibility appealed to here see the entry on 'possibility' in T. Honderich (ed.), *The Oxford Companion to Philosophy*, Oxford: Oxford University Press, 1995, pp. 706–7.

30 See M. Boden, 'Optimism', *Philosophy* 41(158), 1966, 291–303.

31 P. Stratton-Lake, 'The Future of Reason: Kant's Conception of the Finitude of Thinking', PhD thesis, University of Essex, 1990, p. 129.

32 This claim becomes false when circumstances become so dire that there genuinely is no way back from climate change fit to destroy all life on Earth – for example, in a 7°C+ world – or if human nature is incapable of moral action. In such cases, Rawls – citing Kant – seems to me to set the right tone: '[if justice] is not possible, and human beings are largely amoral, if not incurably cynical and self-centred, one might ask, with Kant, whether it is worthwhile for human beings to live on the earth', J. Rawls, *The Law of Peoples*, Cambridge, MA: Harvard University Press, 1999, p. 128.

33 For reflections bearing on this point see C.R. Snyder, 'Hope Theory: Rainbows in the Mind', *Psychological Inquiry* 13(4), 2002, 249–75.

34 Compare Rawls: '[S]o long as we believe for good reasons that a self-sustaining and reasonably just political and social order both at home and abroad is possible,

we can reasonably hope that we or others will someday, somewhere, achieve it; and we can then do something toward this achievement,' Rawls, *The Law of Peoples*, p. 128. Cf. also W. James, *The Will to Believe and Other Essays in Popular Philosophy*, Harvard: Harvard University Press, 1990.

35 See Snyder, 'Hope Theory: Rainbows in the Mind'.

36 I borrow this term from Parfit, but I do not adopt his statement of the principle, and I am not considering it, as he does, as a proposed solution to the 'numbers' problem. See D. Parfit, 'Justifiability to Each Person', *Ratio (New Series)* 16, 2003, 368–90, p. 376.

37 See J. Rawls, *A Theory of Justice*, rev. edn, Oxford: Oxford University Press, 1999.

38 See R. Garnaut, S. Howes, F. Jotzo and P. Sheehan, 'Emissions in the Platinum Age: The Implications of Rapid Development for Climate Change Mitigation', *Oxford Review of Economic Policy* 24(2), 2008, 377–401.

6 Conclusion

1 *Today*, BBC Radio 4, 22 August 2007, 0836hrs.

2 See, for example, *The Stern Review*, which states that '[t]he investments made in the next 10–20 years could lock in very high emissions for the next half-century, or present an opportunity to move the world onto a more sustainable path', Executive Summary, p. xxii.

3 Note the claim is not that the rich would be made *worst off* by global warming, or that its effects are *worse* for those with most than for those with least; clearly, a Bangladeshi who loses everything in a flood fares far worse than a Manhattan resident who loses her condo, car, and wardrobe of designer clothes, even though the former loses less in dollar terms than the latter.

4 I should note that I am aware of stepping into deep waters here: lots of work in psychology, and economics, has been done that addresses this – and related – questions. See, for example, L. Whitmarsh and I. Lorenzoni, 'Perceptions, Behavior and Communication of Climate Change', *WIREs Climate Change*, 1, 2010, 158–61; and M. Hulme, *Why We Disagree about Climate Change: Understanding Controversy, Inaction and Opportunity*, Cambridge: Cambridge University Press, 2009. For policy-related discussion see DEFRA, 'A Framework for Pro-Environmental Behaviours'. Online. HTTP: <http://www.defra.gov.uk/evidence/social/behaviour/documents/behaviours-jan08-report.pdf> (accessed 30 March 2011). In line with the themes of the book, I limit my discussion here just to the resources available to address the motivation from within a Rawlsian framework, which is a topic on which very little has been written.

5 See A. Gosseries, 'Historical Emissions and Free-Riding', *Ethical Perspectives* 11(1), 2004, 36–60, for discussion of the 'parasitism' of this form of exploitation.

6 See S.C. Moser and L. Dilling, 'Toward the Social Tipping Point: Creating a Climate for Change' in S.C. Moser and L. Dilling (eds), *Creating a Climate for Change: Communicating Climate Change and Facilitating Social Change*, Cambridge: Cambridge University Press, 2007.

7 See D. Blumstein and C. Saylan, 'The Failure of Environmental Education (and How To Fix It)', *PLoS Biology* 5(5), 2008. Online. HTTP: < http://www.plosbiology.org/article/info:doi/10.1371/journal.pbio.0050120 > (accessed 10 August 2010). It is worth noting in this context that subjects' beliefs about the likelihood of extreme climate change events changed after being shown the film *The Day After Tomorrow*: they came to believe that such events were *less* likely than they had done before watching the film. See T. Lowe, K. Brown, S. Dessai, M. de Franca Doria, K. Haynes and K. Vincent, 'Does Tomorrow Ever Come? Disaster Narrative and Public Perceptions of Climate Change', *Public Understanding of Science* 15, 2006, 435–57. More generally, as the authors of a key report on climate change by the American

Psychological Association note, 'well-meaning attempts to create urgency about climate change by appealing to fear of disasters or health risks frequently lead to the exact opposite of the desired response: denial, paralysis, apathy, or actions that can create greater risks than the one being mitigated'. J. Swim, S. Clayton, T. Doherty, R. Gifford, G. Howard, J. Reser, P. Stern and E. Weber, *Psychology and Global Climate Change: Addressing a Multi-Faceted Phenomenon and Set of Challenges.* A Report by the American Psychological Association's Task Force on the Interface Between Psychology and Global Climate Change, American Psychological Society, 2009. Online. HTTP: http://www.apa.org/science/about/publications/climate-change.pdf> (accessed 10 August 2010), p. 80, citing S.C. Moser and L. Dilling, 'Making Climate Hot: Communicating the Urgency and Challenge of Global Climate Change', *Environment*, 46, 2004, 32–46.

8 See, for example, W. Dwyer, F. Leeming, M. Cobern, B. Porter and J. Jackson, 'Critical Review of Behavioural Intervention to Preserve the Environment', *Environment and Behaviour* 25, 1993, 274–321; F. Kaiser and U. Fuhrer, 'Ecological Behaviour's Dependency on Different Forms of Knowledge', *Applied Psychology: An International Review*, 52, 2003, 598–613.

9 See J. Rawls, *A Theory of Justice*, Oxford: Oxford University Press, 1971, pp. 255–56.

10 J. Rawls, *Justice as Fairness: A Restatement*, Cambridge, MA: Harvard University Press, 2001, p. 160.

11 J. Rawls, *Political Liberalism*, New York: Columbia University Press, 1993, p. 204. This passage is repeated, almost word for word, in Rawls, *Justice as Fairness*, p. 201.

12 For more detailed discussion see C. McKinnon, *Liberalism and the Defence of Political Constructivism*, Basingstoke: Palgrave, 2002, pp. 14–28.

13 Rawls, *A Theory of Justice*, p. 440. Rawls repeatedly stresses the significance of self-respect throughout his work. See also ibid., pp. 92, 107, 443, 543–45, and Rawls, *Political Liberalism*, pp. 106, 203, 318, 319.

14 See McKinnon, *Liberalism and the Defence of Political Constructivism*, chapter 3. See also T.E. Hill Jnr., *Autonomy and Self-Respect*, Cambridge: Cambridge University Press, 1991, esp. chapters 1 and 2.

15 Rawls, *A Theory of Justice*, p. 319.

16 See Rawls, *Justice as Fairness*, p. 60.

17 For discussion of the 'Uncle Tom' figure see T.E. Hill Jnr., *Autonomy and Self-Respect*, chapter 2.

18 See Rawls, *A Theory of Justice*, p. 444.

19 J. Cohen, 'Democratic Equality', *Ethics* 99, 1989, 727–51, p. 737. See also McKinnon, *Liberalism and the Defence of Political Constructivism*, chapter 5.

20 For more on freedom of association as a social basis of self-respect see C. McKinnon, 'Exclusion Rules and Self-Respect', *Journal of Value Inquiry* 34(4), 2000, 491–505.

21 Rawls, *Justice as Fairness*, p. 200.

22 J. Rawls, *A Theory of Justice*, rev. edn, Oxford: Oxford University Press, 1999, p. 462.

23 Argument for this claim can be found in D. Sachs, 'How to Distinguish Self-Respect from Self-Esteem', *Philosophy and Public Affairs* 10(4), 1981, 346–60.

24 Rawls caused some confusion on this point by mistakenly using the terms 'self-respect' and 'self-esteem' interchangeably in *A Theory of Justice*. This conflation is absent from all the later works.

25 Rawls, *A Theory of Justice* (rev. edn), p. 463.

26 Ibid., p. 386.

27 Ibid., p. 388.

28 See ibid., pp. 389–90.

29 See M. Nussbaum, *Hiding from Humanity: Disgust, Shame and the Law*, Princeton: Princeton University Press, 2004.

30 Rawls, *A Theory of Justice* (rev. edn), p. 390. Nussbaum calls this 'constructive shame' and exempts it from her general critique of shame in politics. See Nussbaum, *Hiding from Humanity*, pp. 211–16.
31 Rawls, *A Theory of Justice* (rev. edn), p. 391.
32 Ibid., p. 391.
33 Think of Michael Corleone seeking redemption in *The Godfather Part III*: 'Just when I thought I was out, they pull me back in.'
34 Susanne C. Moser, 'More Bad News: The Risk of Neglecting Emotional Responses to Climate Change Information', S.C. Moser and L. Dilling (eds), *Creating a Climate for Change*, pp. 64–80, pp. 74–75, citing M.E.P. Seligman, 'Can Happiness be Taught?', *Daedalus* 133(2), 2004, 80–87.

BIBLIOGRAPHY

Abizadeh, A., 'Cooperation, Pervasive Impact, and Coercion: On the Scope (not Site) of Distributive Justice', *Philosophy and Public Affairs* 35(4), 2007, 318–58.

Ackerman, F. and Heinzerling, L., *Priceless: On Knowing the Price of Everything and the Value of Nothing*, New York: The New Press, 1994.

Adger, W.N., Paavola, J., Huq, S., and Mace, M.J., *Fairness in Adaptation to Climate Change*, Cambridge, MA: MIT Press, 2006.

Adler, M.D., 'Corrective Justice and Liability for Global Warming', *University of Pennsylvania Law Review* 155(6), 2007, 1859–67.

Anderson, E., *Value in Ethics and Economics*, Cambridge, MA: Harvard University Press, 1993.

Anderson, E., 'Democracy, Public Policy and Lay Assessments of Scientific Testimony', *Episteme* 8(2), 2011, 144–64.

Anderson, C.A., 'Heat and Violence', *Current Directions in Psychological Science* 10(1), 2001, 33–38.

Anderson, K. and Bows, A., 'Reframing the Climate Change Challenge in the Light of Post-2000 Emission Trends', *Philosophical Transactions of the Royal Society A* 366, 2008, 3863–82.

Arneson, R.J., 'Rational Contractarianism, Corrective Justice, and Tort Law', *Harvard Journal of Law and Public Policy* 15(3), 1992, 889–916.

Baker, R. and Strosberg, M., 'Triage and Equality: An Historical Reassessment of Utilitarian Analyses of Triage', *Kennedy Institute of Ethics Journal* 2(2), 1992, 103–23.

Baranzini, A., Chesney, M. and Morisset, J., 'The Impact of Possible Climate Catastrophes on Global Warming Policy', *Energy Policy* 31, 2003, 691–701.

Barnes, B. and Edge, D., *Science in Context: Readings in the Sociology of Science*, Milton Keynes: Open University Press, 1982.

Baron, M., 'Justifications and Excuses', *Ohio State Journal of Criminal Law* 2, 2005, 387–406.

Barry, B., 'Circumstances of Justice and Future Generations' in R.I. Sikora and B. Barry, (eds), *Obligations to Future Generations*.

—— 'Intergenerational Justice in Energy Policy' in D. MacLean and P.G. Brown (eds) *Energy and the Future*.

—— 'Sustainability and Intergenerational Justice' in A. Dobson (ed.), *Fairness and Futurity: Essays on Environmental Sustainability and Social Justice.*

Barry, J., *Rethinking Green Politics: Nature, Virtue and Progress*, London: Sage Publications, 1999.

Bayles, M., 'Harm to the Unconceived', *Philosophy and Public Affairs* 5, 1976, 292–304.

Beck, C., 'How To Talk to a Climate Sceptic', *A Few Things Ill Considered*, 2008. Online. HTTP: http://scienceblogs.com/illconsidered/2008/07/how_to_talk_to_a_sceptic.php (accessed 10 August 2010).

Beckerman, W. and Pasek, J., *Justice, Posterity and the Environment*, Oxford: Oxford University Press, 2001.

Bell, D., 'How Can Political Liberals be Environmentalists?', *Political Studies* 50(4), 2002, 703–24.

—— 'Liberal Environmental Citizenship', *Environmental Politics* 14(2), 2005, 179–94.

—— 'Political Liberalism and Ecological Justice', *Analyse and Kritik* 28, 2006, 206–22.

Bellamy, D., 'Global Warming? What a Load of Poppycock!', *Daily Mail*, 9 July 2004. Online. HTTP: http://www.junkscience.com/july04/Daily_Mail-Bellamy.htm (accessed 23 July 2010).

Ben-David, J., *The Scientist's Role in Society: A Comparative Study*, Chicago: University of Chicago Press, 1984.

Berlin, I., 'The Pursuit of the Ideal' in I. Berlin, *The Crooked Timber of Humanity*, London: John Murray, 1990.

Blumstein, D. and Saylan, C., 'The Failure of Environmental Education (and How To Fix It)', *PLoS Biology* 5(5), 2008. Online. HTTP: < http://www.plosbiology.org/article/info:doi/10.1371/journal.pbio.0050120> (accessed 10 August 2010).

Boden, M., 'Optimism', *Philosophy* 41(158), 1966, 291–303.

Bookchin, M., *Social Ecology and Communalism*, Oakland, CA: AK Press, 2007.

Broome, J., 'Discounting the Future', *Philosophy and Public Affairs* 23(2), 1994, 128–56.

Brown, D. A., *American Heat*, Lanham, MD: Rowman and Littlefield Publishers Inc., 2002.

Bryson, B., *A Short History of Nearly Everything*, London: Black Swan Books, 2004.

Budescu, D. V., Broomell, S. and Por, H., 'Improving Communication of Uncertainty in the Reports of the Intergovernmental Panel on Climate Change', *Psychological Science*, 20(3), 2009, 299–308.

Calabresi, G.,*The Costs of Accidents: A Legal and Economic Analysis*, New Haven: Yale University Press, 1970.

Caney, S., 'Cosmopolitan Justice, Responsibility and Global Climate Change', *Leiden Journal of International Law* 18(4), 2005, 747–5.

—— 'Global Distributive Justice and the State', *Political Studies* 56, 2008, 487–518.

—— 'Justice and the Distribution of Greenhouse Gas Emissions', *Journal of Global Ethics* 5(2), 2009, 125–46.

Care, N.S., 'Future Generations, Public Policy, and the Motivation Problem', *Environmental Ethics* 4, 1982, 195–213.

Carter, N., *The Politics of the Environment: Ideas, Activism, Policy*, Cambridge: Cambridge University Press, 2001.

Childress, J., 'Triage in Response to a Bioterrorist Attack', in J.D. Moreno (ed.), *In The Wake of Terror: Medicine and Morality in a Time of Crisis.*

Christy, J., Spencer, R.W., Norris, W.B., Braswell, W.D. and Parker, D.E., 'Error Estimates of Version 5.0 of MSU-AMSU Bulk Atmospheric Temperatures', *Journal of Atmospheric Oceanic Technology* 20, 2003, 613–29.

Cohen, J., 'Democratic Equality', *Ethics* 99, 1989, 727–51.
—— 'For a Democratic Society' in S. Freeman (ed.), *The Cambridge Companion to Rawls*.
Coleman, J., 'The Morality of Strict Tort Liability', *William and Mary Law Review* 18, 1976, 259–86.
—— 'Risks and Wrongs', *Harvard Journal of Law and Public Policy* 15(3), 1992, 637–49.
—— *Risks and Wrongs*, Cambridge: Cambridge University Press, 1992.
—— 'Theories of Tort Law' in E.N. Zalta (ed.), *The Stanford Encyclopedia of Philosophy*, 2003. Online. HTTP: <http://plato.stanford.edu/entries/tort-theories/> (accessed 10 August 2010).
—— and S. Shapiro (eds) *The Oxford Handbook of Jurisprudence and Legal Philosophy*, Oxford: Oxford University Press, 2001.
Conquest, R., *The Harvest of Sorrow: Soviet Collectivization and the Terror-Famine*, Oxford: Oxford University Press, 1986.
Coward, H. and Hurka, T. (eds), *Ethics and Climate Change: The Greenhouse Effect*, Waterloo, Ontario: Wilfrid Laurier University Press, 1993.
Cozzens, S.E. and Woodhouse, E.J., 'Science, Government, and the Politics of Knowledge' in S. Jasanoff, G.E. Markle, J.C. Petersen, and T. Pinch (eds), *Handbook of Science and Technology Studies*.
Crocker, D. and Linden, T. (eds), *Ethics of Consumption: The Good Life, Justice, and Global Stewardship*, Lanham, MD: Rowman and Littlefield Publishers, Inc., 1998.
Dark, K., 'Ecological Change and Political Crisis' in B. Holden (ed.), *The Ethical Dimensions of Global Change*.
De-Shalit, A., *Why Posterity Matters: Environmental Policies and Future Generations*, London: Routledge, 1995.
Dessai, S., O'Brien, K. and Hulme, M., 'On Uncertainty and Climate Change', *Global Environmental Change* 17, 2007, 1–3.
Dessler, A.E. and Parson, E.A., *The Science and Politics of Global Climate Change: A Guide to the Debate*, Cambridge: Cambridge University Press, 2006.
Dickens, G., 'The Blast in the Past', *Nature* 401, 1999, 752–55.
Dimitrov, R.S., 'Precaution in Global Environmental Politics', *International Journal of Global Environmental Issues* 5(1/2), 2005, 96–113.
Dobson, A. (ed.), *Fairness and Futurity: Essays on Environmental Sustainability and Social Justice*, Oxford: Oxford University Press, 1999.
—— *Green Political Thought*, 3rd edn, London: Routledge, 2000.
Dryzek, J.S., *The Politics of the Earth: Environmental Discourses*, Oxford: Oxford University Press, 1997.
Dwyer, W., Leeming, F., Cobern, M., Porter, B. and Jackson, J., 'Critical Review of Behavioural Intervention to Preserve the Environment', *Environment and Behaviour* 25, 1993, 274–321.
Eckersley, R., *Environmentalism and Political Theory: Toward an Ecocentric Approach*, London: UCL Press, 1992.
Edwards, P.N., 'Representing the Global Atmosphere: Computer Models, Data, and Knowledge about Climate Change' in C.A. Miller and P.N. Edwards (eds), *Changing the Atmosphere*.
Ekeli, K.S., 'Environmental Risks, Uncertainty and Intergenerational Ethics', *Environmental Values* 13, 2004, 421–48.
Elster, J., *Local Justice: How Institutions Allocate Scarce Goods and Necessary Burdens*, New York: Russell Sage Foundation, 1992.
English, J., 'Justice between Generations', *Philosophical Studies* 31, 1977, 91–104.

European Environment Agency, *The Precautionary Principle in the 20th Century*, London: Earthscan, 2002.

Farber, D.A., 'Basic Compensation for Victims of Climate Change', *University of Pennsylvania Law Review* 155, 2007, 1605–56.

—— 'Tort Law in the Era of Climate Change, Katrina and 9/11: Exploring Liability for Extraordinary Risks', UC Berkeley Public Law Research Paper No. 11211125, 2008. Online. HTTP: http://ssrn.com/abstract=1121125 (accessed 10 August 2010).

—— 'The Case for Climate Compensation: Doing Justice to Climate Change Victims in a Complex World', *Utah Law Review* 2, 2008, 1–36.

—— 'Apportioning Climate Change Costs', *UCLA Journal of Environmental Law and Policy* 21, 2008, 21–54.

Farrelly, C., 'Justice in Ideal Theory: A Refutation', *Political Studies* 55, 2007, 844–64.

Faure, M.G. and Nollkaemper, A., 'International Liability as an Instrument to Prevent and Compensate for Climate Change', *Stanford Journal of International Law* 123, 2007, 124–79.

Finkelstein, C., 'Is Risk a Harm?', *University of Pennsylvania Law Review* 151, 2003, 963–1001.

Fischer, J.M. and Ennis, R.H., 'Causation and Liability', *Philosophy and Public Affairs* 15(1), 1986, 33–40.

Freeman III, A. Myrick, 'Economics' in D. Jamieson (ed.), *A Companion to Environmental Philosophy*.

Freeman, S. (ed.), *John Rawls: Collected Papers*, Cambridge, MA: Harvard University Press, 1999.

—— (ed.), *The Cambridge Companion to Rawls*, Cambridge: Cambridge University Press, 2003.

—— *Justice and the Social Contract*, Oxford: Oxford University Press, 2007.

—— 'The Law of Peoples, Social Cooperation, and Human Rights', in S. Freeman, *Justice and the Social Contract*, Oxford: Oxford University Press, 2007.

—— 'Distributive Justice and the Law of Peoples', in S. Freeman, *Justice and the Social Contract*.

Gardiner, S.M., 'Ethics and Global Climate Change', *Ethics* 114, 2004, 555–600.

—— 'A Core Precautionary Principle', *The Journal of Political Philosophy* 14(1), 2006, 33–60.

—— 'A Perfect Moral Storm: Climate Change, Intergenerational Ethics and the Problem of Corruption', *Environmental Values* 15 (3), 2006, 397–413.

—— 'Saved by Disaster? Abrupt Climate Change, Political Inertia and the Possibility of an Intergenerational Arms Race', *Journal of Social Philosophy* 40(2), 2009, 140–62.

Garnaut, R., Howes, S., Jotzo, F. and Sheehan, P., 'Emissions in the Platinum Age: The Implications of Rapid Development for Climate Change Mitigation', *Oxford Review of Economic Policy* 24(2), 2008, 377–401.

Gaspart, F. and Gosseries, A., 'Are Generational Savings Unjust?', *Politics, Philosophy and Economics* 6, 2007, 193–217.

Gatter, R.A. and Moskop, J.C., 'From Futility to Triage', *The Journal of Medicine and Philosophy* 20, 1995, 191–205.

Gelbspan, R., *The Heat is On: The High Stakes Battle over Earth's Threatened Climate*, Reading, MA: Addison-Wesley Publishing Co., Inc, 1997.

Gieryn, T.F., 'Boundaries of Science' in S. Jasanoff, G.E. Markle, J.C. Petersen, and T. Pinch (eds), *Handbook of Science and Technology Studies*.

Girill, T.R., 'Three Problems Regarding Medical Triage', *Metamedicine* 1, 1980, 135–53.

Godard, O., 'Integrating Scientific Expertise into Regulatory Decision-Making: Social Decision-Making under Conditions of Scientific Controversy Expertise and the Precautionary Principle', EUI Working Paper RSC No. 96/6, 1992.

Goodin, R.E., 'Discounting Discounting', *Journal of Public Policy* 2(1), 1982, 53–72.

—— *Green Political Theory*, Cambridge: Polity Press, 1992.

Gosseries, A., 'Constitutionalizing Future Rights?', *Intergenerational Justice Review* 2, 2004, 10–11.

—— 'Historical Emissions and Free-Riding', *Ethical Perspectives* 11(1), 2004, 36–60.

—— 'Cosmopolitan Luck, Egalitarianism and Climate Change', *Canadian Journal of Philosophy* 31, 2007, 279–309.

—— and Meyer, L.H. (eds), *Intergenerational Justice*, Cambridge: Cambridge University Press, 2009.

Greenpeace UK, 'Palm Oil'. Online. HTTP: <http://www.greenpeace.org.uk/forests/palm-oil> (accessed 12 August 2010).

Griffin, A. and Thomasma, D.C., 'Triage and Critical Care of Children', *Theoretical Medicine* 4, 1983, 155–63.

Gross, M.L., *Bioethics and Armed Conflict: Moral Dilemmas of Medicine and War*, Cambridge, MA: MIT Press, 2006.

Grubb, M., 'The Greenhouse Effect: Negotiating Targets', *International Affairs*, 66(1), 1990, 67–89.

—— 'Seeking Fair Weather: Ethics and the International Debate on Climate Change', *International Affairs* 71(3), 1995, 463–96.

Ha-Duong, M., Swart, R. and Bernstein, L., 'Uncertainty Management in the IPCC', *Global Environmental Change* 17, 2007, 8–11.

Haas, P.M., 'Introduction: Epistemic Communities and International Policy Coordination', *International Organization* 46(1), 1992, 1–35.

Hacking, I., 'Culpable Ignorance of Interference Effects' in D. MacLean (ed.), *Values at Risk*.

Hailwood, S., *How To Be a Green Liberal*, Chesham: Acumen, 2004.

Haller, S., 'A Prudential Argument for Precaution under Uncertainty and High Risk', *Ethics and the Environment* 5(2), 2000, 163–74.

—— *Apocalypse Soon?* Montreal and Kingston: McGill-Queen's University Press, 2002.

Hampson, F.O. and Reppy, J. (eds), *Earthly Goods: Environmental Change and Social Justice*, Ithaca: Cornell University Press, 1996.

Hampton, J., 'Contracts and Choices: Does Rawls Have a Social Contract Theory?', *The Journal of Philosophy* 77(6), 1980, 315–38.

Hansen, J., 'Target Atmospheric CO2: Where Should Humanity Aim?', *Open Journal on Atmospheric Sciences* 2, 2008, 217–31.

—— 'Tipping Point: Perspective of a Climatologist', in E. Fearn and K.H. Redford (eds), *The State of the Wild: A Global Portrait of Wildlife, Wild Lands, and Oceans 2008–9*, Wildlife Conservation Society/Island Press: 2008. Online. HTTP: http://www.columbia.edu/~jeh1/2008/StateOfWild_20080428.pdf (accessed 10 August 2010).

—— 'The Earth's Energy Imbalance: Confirmation and Implications', *Science* 308, 2005, 1431–35. Online. HTTP: http://pubs.giss.nasa.gov/authors/jhansen.html (accessed 5 July 2007).

Hansson, S.O., 'The Limits of Precaution', *Foundations of Science* 2(2), 1997, 293–306.

Hardin, G., 'The Tragedy of the Commons', *Science*, New Series 162(3859), 1968, 1243–48.

Harpwood, V., *Modern Tort Law*, 6th edn, London: Cavendish Publishing Ltd., 2005.

Harremoës, P., Gee, D., MacGarvin, M., Stirling, A., Keys, J., Wynee, B. and GuedesVaz, S. (eds), *The Precautionary Principle in the 20th Century: Late Lessons from Early Warnings*, London: Earthscan, 2002.

Harris, P., 'Bush Covers Up Climate Research', *The Observer*, 21 September 2003. Online. HTTP: <http://www.guardian.co.uk/climatechange/story/0,12374,1046389,00.html> (accessed 23 July 2010).

Harris, P.G., 'Fairness, Responsibility, and Climate Change', *Ethics and International Affairs* 17, 2003, 149–56.

Hayward, T., 'Anthropocentrism: A Misunderstood Problem', *Environmental Values* 6, 1997, 49–63.

Hill Jnr., T.E., *Autonomy and Self-Respect*, Cambridge: Cambridge University Press, 1991.

Holbrook, D., 'The Consequentialistic Side of Environment Ethics', *Environmental Values* 6, 1997, 87–96.

Holden, B. (ed.), *The Ethical Dimensions of Global Change*, Basingstoke: Macmillan, 1996.

—— 'Democratic Theory and the Problem of Global Warming' in B. Holden (ed.), *The Ethical Dimensions of Global Change*, 135–66.

—— *Democracy and Global Warming*, London: Continuum, 2002.

Honderich, T. (ed.), *The Oxford Companion to Philosophy*, Oxford: Oxford University Press, 1995.

Houghton, J., *Global Warming: The Complete Briefing*, 4th edn, Cambridge: Cambridge University Press, 2009.

Houghton, J.T., Ding, Y., Griggs, D.J., Noguer, M., van der Linden, P.J., Dai, X., Maskell, K. and Johnson, C.A. (eds), *Climate Change 2001: The Scientific Basis. Contribution of Working Group I to the Third Assessment Report of the Intergovernmental Panel on Climate Change*, Cambridge: Cambridge University Press, 2001. Online. HTTP: <http://www.grida.no/publications/other/ipcc_tar/ > (accessed 11 August 2010).

Hulme, M., *Why We Disagree about Climate Change: Understanding Controversy, Inaction and Opportunity*, Cambridge: Cambridge University Press, 2009.

Hume, D., *Enquiries Concerning Human Understanding and Concerning the Principles of Morals*, Oxford: Clarendon Press, 1975.

Hunter, D. and Salzman, J., 'Negligence in the Air: The Duty of Care in Climate Change Litigation', *University of Pennsylvania Law Review* 155(6), 2007, 1741–94.

Huntingford, C., Jones, R.G., Prudhomme, C., Lamb, R., Gash, J.H.C. and Jones, D.A., 'Regional Climate-Model Predictions of Extreme Rainfall for a Changing Climate', *Quarterly Journal of the Royal Meteorological Society* 129(590), 2003, 1607–21.

IPCC, 'Guidance Notes for Lead Authors of the IPCC Fourth Assessment Report on Addressing Uncertainties', 2005. Online. HTTP: http://www.ipcc.ch/pdf/supporting-material/uncertainty-guidance-note.pdf (accessed 10 August 2010).

James, W., *The Will to Believe and Other Essays in Popular Philosophy*, Harvard: Harvard University Press, 1990.

Jamieson, D., 'Climate Change and Global Environmental Justice' in C.A. Miller and P.N. Edwards (eds), *Changing the Atmosphere*.

—— (ed.), *A Companion to Environmental Philosophy*, Oxford: Blackwell, 2003.

Jasanoff, S., 'Science and Norms in Global Environmental Regimes' in F.O. Hampson and J. Reppy (eds), *Earthly Goods: Environmental Change and Social Justice*.

—— 'Image and Imagination: The Formation of Global Environmental Consciousness' in C.A. Miller and P.N. Edwards (eds), *Changing the Atmosphere*.

Jasanoff, S., Markle, G.E., Petersen, J.C. and Pinch, T. (eds), *Handbook of Science and Technology Studies*, Thousand Oaks: Sage, 1994.

Kagan, S., 'Causation, Liability, and Internalism', *Philosophy and Public Affairs* 15(1), 1986, 41–59.

Kahn, P.H. *The Human Relationship with Nature*, Cambridge, MA: MIT Press, 1999.

Kaiser, F. and Fuhrer, U., 'Ecological Behaviour's Dependency on Different Forms of Knowledge', *Applied Psychology: An International Review* 52, 2003, 598–613.

Kaiser, H.M. and Drennen, T.E. (eds), *Agricultural Dimensions of Climate Change*, Florida: St. Lucie Press, 1993.

Kavka, G. 'The Paradox of Future Individuals', *Philosophy and Public Affairs* 11, 1982, 93–112.

Kidder, D. and Worsley, T., 'Causes and Consequences of Extreme Permo-Triassic Warming to Globally Equitable Climate and Relation to the Permo-Triassic Extinction and Recovery', *Palaeogeography, Palaeoclimatology, Palaeoecology* 203(3–4), 2004, 207–37.

Krebs, J.and Kacelnik, A., 'Risk: A Scientific View' in *Science, Policy and Risk*, London: The Royal Society, 1997, 31–44.

Kripke, S., *Naming and Necessity*, Oxford: Basil Blackwell, 1972.

Kump, L., Pavlov, A. and Arthur, M.A., 'Massive Release of Hydrogen Sulfide to the Surface Ocean and Atmosphere During Intervals of Ocean Anoxia', *Geology* 33(5), 2005, 367–400.

Landes, W.M., *The Economic Structure of Tort Law*, Harvard: Harvard University Press, 1987.

Larmore, C. 'Public Reason' in S. Freeman (ed.), *The Cambridge Companion to Rawls.*

Laslett, P. and Fishkin, J.S., *Justice between Age Groups and Generations*, New Haven: Yale University Press, 1992.

Law, B., 'Can Palm Oil Help Indonesia's Poor?', *The Guardian*, 1 March 2010. Online. HTTP: <http://news.bbc.co.uk/1/hi/world/asia-pacific/8534031.stm> (accessed 12 August 2010).

Lemons, J., Shrader-Frechette, K. and Cranor, C., 'The Precautionary Principle: Scientific Uncertainty and Type I and Type II Errors', *Foundations of Science* 2(2), 1997, 207–36.

Lewens, T. (ed.), *Risk: Philosophical Perspectives*, London: Routledge 2007.

Lewenstein, B.V., 'Science and the Media' in S. Jasanoff, G.E. Markle, J.C. Petersen, and T. Pinch, T. (eds), *Handbook of Science and Technology Studies.*

Lomborg, B., *The Skeptical Environmentalist: Measuring the Real State of the World*, Cambridge: Cambridge University Press, 2001.

Lowe, T., Brown, K., Dessai, S., de Franca Doria, M., Haynes, K. and Vincent, K., 'Does Tomorrow ever Come? Disaster Narrative and Public Perceptions of Climate Change', *Public Understanding of Science* 15, 2006, 435–57.

Lynas, M., *High Tide: How Climate Crisis Is Engulfing Our Planet*, London: Harper Perennial, 2005.

——, *Six Degrees: Our Future on a Hotter Planet*, London: HarperCollins, 2007.

McCarthy, C., *The Road*, New York: Vintage Books, 2006.

McCarthy, D., 'Liability and Risk', *Philosophy and Public Affairs* 25(3), 1996, 238–62.

McGarvin, M. 'Science, Precaution, Facts and Value' in T. O'Riordan, J. Cameron and A. Jordan (eds), *Reinterpreting the Precautionary Principle.*

McGuire, B., *Global Catastrophes: A Very Short Introduction*, Oxford: Oxford University Press, 2005.

McKinnon, C., 'Exclusion Rules and Self-Respect', *Journal of Value Inquiry*, 34(4), 2000, 491–505.

—— *Liberalism and the Defence of Political Constructivism*, Basingstoke: Palgrave Macmillan, 2002.

—— *Toleration: A Critical Introduction*, London: Routledge, 2005.

—— 'Cosmopolitan Hope' in H. Brighouse and G. Brock (eds), *The Political Philosophy of Cosmopolitanism*, Cambridge: Cambridge University Press, 2005.

—— 'Should We Tolerate Holocaust Denial?', *Res Publica* 12(4), 2007, 9–28.

MacLean, D. and Brown, P.G. (eds), *Energy and the Future*, New Jersey: Rowman and Allenheld, 1983.

Makohon, P., 'Witness: Memoirs of the Famine of 1933 in Ukraine'. Online. HTTP: <http://faminegenocide.com/resources/witness/main.html> (accessed 27 July 2010).

Manabe, S., Wetherald, R.T., Milly, P.C.D., Delworth, T.L. and Stouffer, R.J., 'Century-scale Change in Water Availability: CO2 Quadrupling Experiment', *Climatic Change* 64, 2004, 59–76.

Mann, M.E., Bradley, R.S. and Hughes, M.K., 'Northern Hemisphere Temperatures during the Past Millennium: Inferences, Uncertainties, and Limitations', *Geophysical Research Letters* 26(6), 1999, 759–62. Online. HTTP: < http://www.ltrr.arizona.edu/webhome/aprilc/data/my%20stuff/MBH1999.pdf > (accessed 19 August 2010).

Manning, M., Petit, M., Easterling, D., Murphy, J., Patwardhan, A., Rogner, H., Swart, R. and Yohe, G., 'Describing Scientific Uncertainties in Climate Change to Support Analysis of Risk and of Options', Report on IPCC Workshop on Uncertainty and Risk, 2004. Online. HTTP: http://www.ipcc.ch/pdf/supporting-material/ipcc-workshop-2004-may.pdf (accessed 10 August 2010).

Marohasy, J., 'Environmental Fundamentalism', *Policy* 20(3), 2004.

Met Office, 'Avoiding Dangerous Climate Change', Met Office, 2008. Online. HTTP: http://www.metoffice.gov.uk/publications/brochures/cop14.pdf> (accessed 10 August 2010).

Metz, B., Davidson, O.R., Bosch, P.R., Dave, R. and Meyer L.A. (eds), *Climate Change 2007: Working Group III: Mitigation of Climate Change*, Cambridge: Cambridge University Press, Cambridge, 2007. Online. HTTP: < http://www.ipcc.ch/publications_and_data/ar4/wg3/en/contents.html > (accessed 10 August 2010).

Meyer, L. and Roser, D., 'Distributive Justice and Climate Change. The Allocation of Emission Rights', *Analyse and Kritik* 28, 2006, 223–49.

Mill, J.S., *On Liberty*, Cambridge: Cambridge University Press, 1989.

Miller, C.A., 'Challenges in the Application of Science to Global Affairs: Contingency, Trust, and Moral Order' in C.A. Miller and P.N. Edwards (eds), *Changing the Atmosphere*.

—— and Edwards, P.N., *Changing the Atmosphere: Expert Knowledge and Environmental Governance*, Cambridge, MA: MIT Press, 2001.

Miller, D., 'Social Justice and Environmental Goods' in A. Dobson (ed.), *Fairness and Futurity: Essays on Environmental Sustainability and Social Justice*.

Monbiot, G., 'Junk Science', *The Guardian*, 10 May 2005. Online. HTTP: http://www.monbiot.com/archives/2005/05/10/junk-science/ (accessed 23 July 2010).

——, *Heat: How To Stop the Planet Burning*, London: Allen Lane, 2006.

Moreno, J.D. (ed.), *In the Wake of Terror: Medicine and Morality in a Time of Crisis*, Cambridge, MA: MIT Press, 2003.

Moser, S.C., 'More Bad News: The Risk of Neglecting Emotional Responses to Climate Change Information' in S.C. Moser and L. Dilling (eds), *Creating a Climate for Change: Communicating Climate Change and Facilitating Social Change*.

—— 'Making Climate Hot: Communicating the Urgency and Challenge of Global Climate Change', *Environment* 46, 2004, 32–46.

—— 'Toward the Social Tipping Point: Creating a Climate for Change' in S.C. Moser and L. Dilling (eds), *Creating a Climate for Change: Communicating Climate Change and Facilitating Social Change*.

—— and Dilling, L. (eds), *Creating a Climate for Change: Communicating Climate Change and Facilitating Social Change*, Cambridge: Cambridge University Press, 2007.

Motavalli, J. (ed.), *Feeling the Heat: Dispatches from the Frontlines of Climate Change*, New York: Routledge, 2004.
Munasinghe, M. and Swart, R., *Primer on Climate Change and Sustainable Development: Facts, Policy Analysis and Applications*, Cambridge: Cambridge University Press, 2005.
Murphy, L.B., *Moral Demands in Nonideal Theory*, Oxford: Oxford University Press, 2000.
Nagel, T., 'Moral Luck' in his *Mortal Questions*.
—— *Mortal Questions*, Cambridge: Cambridge University Press, 1979.
Nakicenovic, N. and Swart, R. *Special Report on Emissions Scenarios*, Cambridge: Cambridge University Press, 2000. Online. HTTP: < http://www.ipcc.ch/ipccreports/sres/emission/index.php?idp=0 > (accessed 10 August 2010).
Nielsen, J.K. and Shen, Y., 'Evidence for Sulphidic Deep Water during the Late Permian in the East Greenland Basin', *Geology* 32(12), 2004, 1037–40.
Nussbaum, M., *Hiding from Humanity: Disgust, Shame and the Law*, Princeton: Princeton University Press, 2004.
O'Neill, O., 'Necessary Anthropocentrism and Contingent Speciesism', *Proceedings of the Aristotelian Society* Supplementary Vol. 72, 1998, 211–28.
Oreskes, N., 'The Scientific Consensus on Climate Change', *Science* 306 (5702), 2004, 1686–87.
O'Riordan, T., 'The Precautionary Principle and Civic Science' in T. O'Riordan, J. Cameron, and A. Jordan (eds), *Reinterpreting the Precautionary Principle*.
——, Cameron, J. and Jordan, A. (eds), *Reinterpreting the Precautionary Principle*, London: Cameron May Ltd., 2001.
Otsuka, M., 'Moral Luck: Optional not Brute', *Philosophical Perspectives* 23, 2009, 373–88.
Owen, D.G. (ed.), *Philosophical Foundations of Tort Law*, Oxford: Clarendon Press, 1995.
—— 'Why Philosophy Matters to Tort Law' in D.G. Owen (ed.) *Philosophical Foundations of Tort Law*.
Pachauri, R.K and Reisinger, A. (eds), *Climate Change 2007: Synthesis Report. Contribution of Working Groups I, II and III to the Fourth Assessment Report of the Intergovernmental Panel on Climate Change*, Geneva: IPCC, 2007. Online. HTTP: < http://www.ipcc.ch/publications_and_data/ar4/syr/en/contents.html > (accessed 10 August 2010).
Paden, R., 'Rawls' Just Savings Principle and the Sense of Justice', *Social Theory and Practice* 23(1), 1997, 27–51.
Page, E., 'Intergenerational Justice and Climate Change', *Political Studies* 47, 1999, 53–66.
—— *Climate Change, Justice and Future Generations*, Cheltenham, UK: Edward Elgar, 2006.
Palin, S., 'Sarah Palin on the Politicization of the Copenhagen Climate Conference', *The Washington Post*, 9 December 2009. Online. HTTP: http://www.washingtonpost.com/wp-dyn/content/article/2009/12/08/AR2009120803402.html. (accessed 26 July 2010).
Parfit, D., 'Innumerate Ethics', *Philosophy and Public Affairs* 7(4), 1978, 285–301.
—— 'Energy Policy and the Further Future: The Identity Problem' in D. MacLean and P.G. Brown (eds) *Energy and the Future*.
—— *Reasons and Persons*, Oxford: Clarendon Press, 1984.
—— 'Justifiability to Each Person', *Ratio (New Series)* 16, 2003, 368–90.
Parry, M.L., Canziani, O.F., Palutikof, J.P., van der Linden, P.J. and Hanson, C.E. (eds), *Contribution of Working Group II to the Fourth Assessment Report of the Intergovernmental Panel on Climate Change*, Cambridge: Cambridge University Press, 2007. Online. HTTP: <http://www.ipcc.ch/publications_and_data/publications_ipcc_fourth_assessment_report_wg2_report_impacts_adaptation_and_vulnerability.htm> (accessed 10 August 2010).

Pearce, F., *With Speed and Violence: Why Scientists Fear Tipping Points in Climate Change*, Boston: Beacon Press, 2007.

—— 'How the "Climategate" Scandal is Bogus and Based on Climate Sceptics' Lies', *The Guardian*, 9 February 2010. Online. HTTP: <http://www.guardian.co.uk/environment/2010/feb/09/climategate-bogus-sceptics-lies (accessed 11 August 2010).

—— 'Climate Wars: Guardian Special Investigation', 2010. Online. HTTP: http://www.guardian.co.uk/environment/series/climate-wars-hacked-emails (accessed 11 August 2010).

Perry, S.R., 'The Mixed Conception of Corrective Justice', *Harvard Journal of Law and Public Policy* 15(3), 1992, 917–39.

—— 'Risk, Harm, Interests and Rights', in T. Lewens (ed.), *Risk: Philosophical Perspectives*.

Pew Research Center, *Luxury or Necessity? Things We Can't Live without: The List Has Grown in the Past Decade*, Pew Research Center: A Social Trends Report, 2006.

Pollack, H.N. *Uncertain Science … Uncertain World*, Cambridge: Cambridge University Press, 2003.

Pope, V., 'The Scientific Evidence for Early Action on Climate Change', Met Office. Online. HTTP: < http://www.metoffice.gov.uk/climatechange/policymakers/action/evidence.html > (accessed 10 August 2010).

Posner, E. and Sunstein, C.R., 'Climate Change Justice', John M. Olin Law and Economics Working Paper No. 354 (2D series), Public Law and Legal Theory Working Paper No. 177. Online. HTTP: <http://papers.ssrn.com/sol3/papers.cfm?abstract_id=1008958> (accessed 10 August 2010).

Posner, R.A., *Catastrophe: Risk and Response*, Oxford: Oxford University Press, 2004.

Postema, G.J. 'Risks, Wrongs, and Responsibility: Coleman's Liberal Theory of Commutative Justice', *Yale Law Journal* 103, 1993–94, 861–97.

—— (ed.), *Philosophy and the Law of Torts*, Cambridge: Cambridge University Press, 2001.

Raffernsperger, C., 'Uses of the Precautionary Principle in International Treaties and Agreements'. Online. HTTP: http://www.biotech-info.net/treaties_and_agreements.html (accessed 21 March 2006).

Rawls, J., *A Theory of Justice*, Oxford: Oxford University Press, 1971.

—— *Political Liberalism*, New York: Columbia University Press, 1993.

—— *A Theory of Justice*, rev. edn, Oxford: Oxford University Press, 1999.

—— *The Law of Peoples*, Cambridge, MA: Harvard University Press, 1999.

—— 'The Idea of Public Reason Revisited' in S. Freeman (ed.), *John Rawls: Collected Papers*.

—— 'Justice as Fairness: Political not Metaphysical' in S. Freeman (ed.), *John Rawls: Collected Papers*.

—— *Justice as Fairness: A Restatement*, Cambridge, MA: Harvard University Press, 2001.

Raz, J., *The Morality of Freedom*, Oxford: Clarendon Press, 1986.

—— 'Facing Diversity: The Case of Epistemic Abstinence', *Philosophy and Public Affairs* 19, 1990, 3–46.

Reiman, J., 'Being Fair to Future People: The Non-Identity Problem in the Original Position', *Philosophy and Public Affairs* 35, 2007, 69–92.

Ripstein, A., *Equality, Responsibility and the Law*, Cambridge: Cambridge University Press, 1999.

—— 'Philosophy of Tort Law' in J.L. Coleman and S. Shapiro (eds), *The Oxford Handbook of Jurisprudence and Legal Philosophy*.

—— 'The Division of Responsibility and the Law of Tort', *Fordham Law Review* 72, 2004, 101–32.

Robeyns, I. 'Ideal Theory in Theory and Practice', *Social Theory and Practice*, 34(3), 2008, 341–62.
Ronzoni, M., 'The Global Order: A Case of Background Injustice? A Practice-Dependent Account', *Philosophy and Public Affairs* 37(3), 2009, 229–56.
Rosenblum, N., *Membership and Morals*, Princeton, NJ: Princeton University Press, 2000.
Rosenthal, E., 'Third-World Stove Soot Is Target in Climate Fight', *New York Times*, 15 April 2009. Online. HTTP: < http://www.nytimes.com/2009/04/16/science/earth/16degrees.html > (accessed 13 August 2010).
Russell, M., 'The Independent Climate Change E-Mails Review', 2010. Online. HTTP: http://www.guardian.co.uk/environment/2010/jul/07/findings-muir-russell-review (accessed 26 July 2010).
Ryskin, G., 'Methane-Driven Oceanic Eruptions and Mass Extinctions', *Geology* 31(9), 2003, 741–44.
Sachs, D., 'How to Distinguish Self-Respect from Self-Esteem'. *Philosophy and Public Affairs* 10(4), 1981, 346–60.
Sandin, P., 'The Precautionary Principle and the Concept of Precaution', *Environmental Values* 13, 2004, 461–75.
Schroeder, C.H., 'Rights against Risks', *Columbia Law Review* 86, 1986, 495–562.
—— 'Corrective Justice and Liability for Increasing Risks', *UCLA Law Review* 37, 1989–90, 439–78.
—— 'Corrective Justice, Liability for Risks, and Tort Law', *UCLA Law Review* 38, 1990–99, 143–61.
—— 'Causation, Compensation, and Moral Responsibility' in D.G. Owen (ed.), *Philosophical Foundations of Tort Law*.
—— 'Prophets, Priests, and Pragmatists', *Minnesota Law Review* 8(7), 2003, 1065–98.
Scmidtz, D., 'A Place for Cost-Benefit Analysis', *Philosophical Issues* 11, 2001, 148–71.
Seligman, M.E.P., 'Can Happiness Be Taught?', *Daedalus* 133(2), 2004, 80–87.
Shapiro, I. and Wagner DeCew, J. (eds), *Theory and Practice: Nomos XXXVII*, New York: New York University Press, 1995.
Sherwin, E., 'Why Is Corrective Justice Just?', *Harvard Journal of Law and Public Policy* 15(3), 1992, 839–49.
Shue, H., 'Four Questions of Justice' in H.M. Kaiser and T.E. Drennen (eds), *Agricultural Dimensions of Climate Change*.
—— 'Subsistence Emissions and Luxury Emissions', *Law and Policy* 15(1), 1993, 39–59.
—— 'After You: May Action by the Rich Be Contingent Upon Action by the Poor?', *Indiana Journal of Global Legal Studies* 1(2), 1994, 343–66.
—— 'Ethics, the Environment and the Changing International Order', *International Affairs* 71(3), 1995, 453–61.
—— 'Environmental Change and the Varieties of Justice' in F.O. Hampson and J. Reppy (eds), *Earthly Goods: Environmental Change and Social Justice*.
—— 'Avoidable Necessity: Global Warming, International Fairness, and Alternative Energy' in I. Shapiro and J. Wagner DeCew (eds), *Theory and Practice: Nomos XXXVII*.
—— 'Climate' in D. Jamieson (ed.), *A Companion to Environmental Philosophy*.
Sikora, R.I. and Barry, B. (eds), *Obligations to Future Generations*, Philadelphia: Temple University Press, 1978.
Singer, P., 'Ethics and Climate Change: A Commentary on MacCracken, Toman and Gardiner', *Environmental Values* 13, 2004, 415–22.
Slovic, P., 'Perception of Risk', *Science*, New Series 236(4799), 1987, 280–85.

Snyder, C.R. 'Hope Theory: Rainbows in the Mind', *Psychological Inquiry* 13(4), 2002, pp. 249–75.

Solomon, S., Qin, D., Manning, M., Chen, Z., Marquis, M., Averyt, K.B., Tignor, M. and Miller, H.L. (eds), *Climate Change 2007: The Physical Science Basis. Contribution of Working Group I to the Fourth Assessment Report of the Intergovernmental Panel on Climate Change*, Cambridge: Cambridge University Press, 2007. Online. HTTP: <http://www.ipcc.ch/publications_and_data/ar4/wg1/en/contents.html> (accessed 10 August 2010).

Solum, L.B., 'To Our Children's Children's Children: The Problems of Intergenerational Ethics', *Loyola of Los Angeles Law Review* 35(1), 2001, 163–233.

Soule, E., 'Assessing the Precautionary Principle', *Public Affairs Quarterly* 14(4), 2000, 309–28.

Spash, C., *Greenhouse Economics: Value and Ethics*, London: Routledge, 2002.

Steiner, H., 'The Rights of Future Generations' in D. MacLean and P.G. Brown (eds) *Energy and the Future.*

Stern, N., Cabinet Office, HM Treasury, *The Economics of Climate Change: The Stern Review*, Cambridge: Cambridge University Press, 2007.

Stratton-Lake, P., 'The Future of Reason: Kant's Conception of the Finitude of Thinking', PhD thesis, University of Essex, 1990.

Sunstein, C., *Laws of Fear: Beyond the Precautionary Principle*, Cambridge: Cambridge University Press, 2005.

—— 'Cost-Benefit Analysis and the Environment', John M. Olin Law and Economics Working Paper No. 227 (2D Series), October 2004, The Law School, University of Chicago. Online. HTTP: http://www.law.uchicago.edu/files/files/227-crs-environment.pdf (accessed 10 August 2010).

—— 'Cost-Benefit Analysis and the Environment', *Ethics* 115, 2005, 351–85.

—— 'The Catastrophic Harm Precautionary Principle', *Issues in Legal Scholarship*, Article 3, 2007. Online. HTTP: http://www.bepress.com/ils/iss10/art3 (accessed 10 August 2010).

—— *Worst Case Scenarios*, Harvard: Harvard University Press, 2007.

Swift, A., 'The Value of Philosophy in Nonideal Circumstances', *Social Theory and Practice* 34(3), 2008, 363–87.

Swim, J., Clayton, S., Doherty, T., Gifford, R., Howard, G., Reser, J., Stern, P. and Weber, E., *Psychology and Global Climate Change: Addressing a Multi-Faceted Phenomenon and Set of Challenges. A Report by the American Psychological Association's Task Force on the Interface between Psychology and Global Climate Change*, American Psychological Society, 2009. Online. HTTP: http://www.apa.org/science/about/publications/climate-change.pdf> (accessed 10 August 2010).

Taurek, J.M., 'Should the Numbers Count?', *Philosophy and Public Affairs* 6(4), 1977, 293–316.

Temkin, L., *Inequality*, Oxford: Oxford University Press, 1993.

Thompson, J., *Intergenerational Justice: Rights and Responsibilities in an Intergenerational Polity*, New York: Routledge, 2009.

Thomson, J.J., 'Remarks on Causation and Liability', *Philosophy and Public Affairs* 13(2), 1984, 101–33.

—— *Rights, Restitution, and Risk: Essays in Moral Theory*, Cambridge, MA: Harvard University Press, 1986.

—— 'A Note on Internalism', *Philosophy and Public Affairs* 15(1), 1986, 60–66.

—— 'Imposing Risks' in J.J. Thomson, *Rights, Restitution, and Risk: Essays in Moral Theory.*

—— 'Some Questions about Government Regulation of Behaviour' in J.J. Thomson, *Rights, Restitution, and Risk: Essays in Moral Theory*.

Turner, D. and Hartzell, L., 'The Lack of Clarity in the Precautionary Principle', *Environmental Values* 13, 2004, 449–60.

Vanderheiden, S., *Atmospheric Justice*, Oxford: Oxford University Press, 2008.

Waldron, J., *Liberal Rights: Collected Papers 1981–1991*, Cambridge: Cambridge University Press, 1993.

—— 'Moments of Carelessness and Massive Loss' in J. Waldron, *Liberal Rights*.

Wall, S., 'Just savings and the difference principle', *Philosophical Studies* 116, 2003, 79–102.

Ward, B., 'Has the Impact of Climategate Worn Off?', *New Scientist*, 14 February 2011. Online. HTTP: < http://www.newscientist.com/blogs/thesword/2011/02/has-the-impact-of-climategate.html > (accessed 21 March 2011).

Weinrib, E.J., 'Correlativity, Personality, and the Emerging Consensus on Corrective Justice', *Theoretical Inquiries in Law (Online Edition)* 2(1), 2001, Article 4. Online. HTTP: <http://www.bepress.com/til/default/vol2/iss1/art4> (accessed 10 August 2010).

Whitmarsh, L. and Lorenzoni, I., 'Perceptions, Behavior and Communication of Climate Change', *WIREs Climate Change* 1, 2010, 158–61.

Williams, B., *Moral Luck*, Cambridge: Cambridge University Press, 1981.

—— 'Moral Luck' in B. Williams, *Moral Luck*.

Wing, S.L., Harrington, G.J., Smith, F.A., Bloch, J.I., Boyer, D.M. and Freeman, K.H., 'Transient Floral Change and Rapid Global Warming at the Paleocene-Eocene Boundary', *Science* 310, 2005, 333–36.

Winslow, G.R., *Triage and Justice*, Berkeley: University of California Press, 1982.

Wissenburg, M., *Green Liberalism: The Free and the Green Society*, London: UCL Press, 1998.

Wolff, J. and de-Shalit, A., *Disadvantage*, Oxford: Oxford University Press, 2007.

Wynne, B., 'Public Understanding of Science' in S. Jasanoff, G.E. Markle, J.C. Petersen, and T. Pinch (eds), *Handbook of Science and Technology Studies*.

Yearley, S., 'The Environmental Challenge to Science Studies' in S. Jasanoff, G.E. Markle, J.C. Petersen, and T. Pinch (eds), *Handbook of Science and Technology Studies*.

Zhao, X., and Belk, R.W., 'Politicizing Consumer Culture: Advertising's Appropriation of Political Ideology in China's Social Transition', *Journal of Consumer Research* 35, 2008, 231–44.

INDEX